Homogeneous Catalysis and Mechanisms in Water and Biphasic Media

Homogeneous Catalysis and Mechanisms in Water and Biphasic Media

Special Issue Editor

Luca Gonsalvi

MDPI • Basel • Beijing • Wuhan • Barcelona • Belgrade

Special Issue Editor
Luca Gonsalvi
Institute of Chemistry of Organometallic Compounds
National Research Council of Italy (ICCOM-CNR)
Italy

Editorial Office
MDPI
St. Alban-Anlage 66
4052 Basel, Switzerland

This is a reprint of articles from the Special Issue published online in the open access journal *Catalysts* (ISSN 2073-4344) from 2016 to 2018 (available at: https://www.mdpi.com/journal/catalysts/special_issues/water_biphasic).

For citation purposes, cite each article independently as indicated on the article page online and as indicated below:

LastName, A.A.; LastName, B.B.; LastName, C.C. Article Title. *Journal Name* **Year**, *Article Number*, Page Range.

ISBN 978-3-03897-584-7 (Pbk)
ISBN 978-3-03897-585-4 (PDF)

Cover image courtesy of pixabay.com user HG-Fotografie.

Contents

About the Special Issue Editor

Luca Gonsalvi (born in Parma, Italy, in 1968) obtained his Laurea Degree in Chemistry (equiv. M. Sc.) in 1994 at the University of Parma (Italy). From 1996 to 1999, he was a member of the group of Dr. Anthony Haynes and Prof. Peter M. Maitlis at the University of Sheffield (UK) as Ph. D. student, with a project on organometallic mechanistic studies on Rh and Ir-catalysed methanol carbonylation, in collaboration with BP Chemicals. In December 1999, he moved to Delft University of Technology (NL) as Postdoctoral Research Associate with Prof. Roger A. Sheldon and Prof. Isabel W. C. E. Arends, developing new catalytic approaches to selective ether oxidation to esters. In December 2001, he joined ICCOM CNR Florence (Italy) as Staff Researcher, working on water and biphasic homogeneous catalysis—the synthesis of water-soluble ligands and organometallic complexes—applied especially to hydrogen and carbon dioxide activation. Since 2010, he has been Senior Researcher and Group Leader at ICCOM CNR. He was awarded the University of Sheffield Turner Prize for Best Ph. D. Chemistry Thesis (2000) and the CNR Career Development Prize (2005). In 2016, he received the Italian National Habilitation as Full Professor in General and Inorganic Chemistry. He has co-authored more than 100 articles on peer-reviewed journals, filed one patent, and presented his work at more than 140 conferences and symposia, including Keynote and Plenary Lectures. He has been Peer Reviewer for 45 scientific journals since 2004. Recently, he has been Scientific Secretary of the 28th International Conference on Organometallic Chemistry (ICOMC 2018) that was held in Florence, Italy, 15–20 July 2018, attended by more than 950 participants from all continents. His current research interests, centred on homogenous catalysis, include organometallic chemistry and catalysis in water, carbon dioxide activation and utilisation, and hydrogen activation by precious and earth-abundant transition metals. Personal Webpage: www.iccom.cnr.it/gonsalvi

Editorial

Homogeneous Catalysis and Mechanisms in Water and Biphasic Media

Luca Gonsalvi

Institute of Chemistry of Organometallic Compounds, National Research Council of Italy (ICCOM-CNR), Via Madonna del Piano 10, 50019 Sesto Fiorentino, Florence, Italy; l.gonsalvi@iccom.cnr.it

Received: 7 November 2018; Accepted: 12 November 2018; Published: 14 November 2018

After its discovery in the early 1980s and successful application on an industrial scale (Ruhrchemie/Rhone-Poulenc process) [1–4], water phase and biphasic catalysis have been the subject of fundamental studies in a relatively limited number of research laboratories around the world [5], almost at a curiosity level. During the last 15 years, however, this topic has witnessed a true renaissance, mainly due to the increased attention of industry and academia to more environmentally friendly processes. Water is the green solvent par excellence, and a great deal of research has been carried out to convey the properties of known transition metal catalysts to their water-soluble analogs, maintaining high activity and selectivity [6]. The keys to success have been, among others, the discovery of synthetic pathways to novel molecular metal-based catalysts [7], new mechanistic insights into the role of water as a non-innocent solvent [8], the identification of reaction pathways through experimental and theoretical methods, the application of novel concepts for phase transfer agents in biphasic catalysis and advances in engineering and related techniques applied to various reactions carried out in aqueous media.

Some of the approaches currently used to tackle these problems are described in the present Special Issue, that collects three review articles and six original research papers. The main cutting-edge approaches developed in the field of aqueous biphasic catalysis using cyclodextrins as a supramolecular tool [9] are discussed and compared in the first review [10]. In the second review [11], the topic of the metal-catalyzed addition of carboxylic acids to alkynes [12,13] as a tool for the synthesis of carboxylate-functionalized olefinic compounds is reviewed, with an emphasis on processes run in water. The synthesis of β-oxo esters by the catalytic addition of carboxylic acids to terminal propargylic alcohols in water is also discussed. The third review article [14] describes the use of an advanced analytical method, high-resolution ultrasonic spectroscopy [15], for the non-destructive real-time monitoring of chemical reactions in complex systems such as emulsions, suspensions and gels. This method has the advantage of being applicable to the monitoring of reactions in continuous media and in micro/nano bioreactors (e.g., nanodroplets of microemulsions), enabling measurements of concentrations of substrates and products over the whole course of reaction, evaluation of kinetic mechanisms, and the measurement of kinetic and equilibrium constants and reaction Gibbs energy.

Two research articles [16,17] describe the use of water-soluble Ru(II) complexes [18] for reactions such as C=C and C=N bond transfer hydrogenation [19], and how to minimize the production of CO during HCOOH dehydrogenation reactions in water media, respectively [20]. Other articles describe applications in speciality reactions and materials, for example the use of chitosan aerogel-catalyzed asymmetric aldol reaction of ketones with isatins in the presence of water [21], the use of bismuth oxyhalide as an activator of peroxide for water purification to degrade carbamazepines [22], the study of catalytic activities of nucleic acid enzymes in dilute aqueous solutions [23], the use of iminodiacetic acid-modified Nieuwland catalysts [24] for acetylene dimerization, and the selective conversion of acetylene to monovinylacetylene (MVA) [25].

In summary, this Special Issue provides an uncommon and multifocal point of view on different fields of application where water can be used as a green solvent and/or has implications in the reaction mechanism, the engineering of a process or an analytical technique. These readings can be of interest and help to colleagues working in related research areas, and stimulate the curiosity of others who may think of water processes as viable—albeit sometimes more difficult—alternatives to traditional approaches.

Finally, I would like to express my deepest gratitude to all authors for their valuable contributions that made this book possible.

Conflicts of Interest: The author declares no conflict of interest.

References

1. Cornils, B.; Kuntz, E.G. Introducing TPPTS and related ligands for industrial biphasic processes. *J. Organomet. Chem.* **1987**, *570*, 177–186. [CrossRef]
2. Kuntz, E. Rhone-Poulenc Recherche. FR Patent 2.314.910, 1975.
3. Kalck, P.; Monteil, F. Use of Water-Soluble Ligands in Homogeneous Catalysis. *Adv. Organomet. Chem.* **1992**, *34*, 219–284.
4. Cornils, B.; Hibbel, J.; Konkol, W.; Lieder, B.; Much, J.; Schmidt, V.; Wiebus, E. (Ruhrchemie AG). EP 0.103.810, 1982.
5. Joó, F. *Aqueous Organometallic Catalysis*, 1st ed.; Springer: Dordrecht, The Netherlands, 2001.
6. Cornils, B.; Herrmann, W.A. (Eds.) *Aqueous-Phase Organometallic Catalysis: Concepts and Applications*, 2nd ed.; Wiley-VCH Verlag GmbH: Weinheim, Germany, 2004.
7. Shaughnessy, K.H. Hydrophilic ligands and their application in aqueous-phase metal-catalyzed reactions. *Chem. Rev.* **2009**, *109*, 643–710. [CrossRef] [PubMed]
8. Dixneuf, P.H.; Cadierno, V. (Eds.) *Metal-Catalyzed Reactions in Water*; Wiley-VCH Verlag GmbH: Weinheim, Germany, 2013.
9. Machut, C.; Patrigeon, J.; Tilloy, S.; Bricout, H.; Hapiot, F.; Monflier, E. Self-assembled supramolecular bidentate ligands for aqueous organometallic catalysis. *Angew. Chem. Int. Ed.* **2007**, *46*, 3040–3042. [CrossRef] [PubMed]
10. Hapiot, F.; Monflier, E. Unconventional Approaches Involving Cyclodextrin-Based, Self-Assembly-Driven Processes for the Conversion of Organic Substrates in Aqueous Biphasic Catalysis. *Catalysts* **2017**, *7*, 173. [CrossRef]
11. Francos, J.; Cadierno, V. Metal-Catalyzed Intra- and Intermolecular Addition of Carboxylic Acids to Alkynes in Aqueous Media: A Review. *Catalysts* **2017**, *7*, 328. [CrossRef]
12. Alonso, F.; Beletskaya, I.P.; Yus, M. Transition-metal-catalyzed addition of heteroatom-hydrogen bonds to alkynes. *Chem. Rev.* **2004**, *104*, 3079–3159. [CrossRef] [PubMed]
13. Beller, M.; Seayad, J.; Tillack, A.; Jiao, H. Catalytic Markovnikov and anti-Markovnikov functionalization of alkenes and alkynes: Recent developments and trends. *Angew. Chem. Int. Ed.* **2004**, *43*, 3368–3398. [CrossRef] [PubMed]
14. Buckin, V.; Altas, M.C. Ultrasonic Monitoring of Biocatalysis in Solutions and Complex Dispersions. *Catalysts* **2017**, *7*, 336. [CrossRef]
15. Buckin, V. Application of High-Resolution Ultrasonic Spectroscopy for analysis of complex formulations. Compressibility of solutes and solute particles in liquid mixtures. *IOP Conf. Ser. Mater. Sci. Eng.* **2012**, *42*, 1–18. [CrossRef]
16. Guerriero, A.; Peruzzini, M.; Gonsalvi, L. Ruthenium(II)-Arene Complexes of the Water-Soluble Ligand CAP as Catalysts for Homogeneous Transfer Hydrogenations in Aqueous Phase. *Catalysts* **2018**, *8*, 88. [CrossRef]
17. Henricks, V.; Yuranov, I.; Autissier, N.; Laurenczy, G. Dehydrogenation of Formic Acid over a Homogeneous Ru-TPPTS Catalyst: Unwanted CO Production and Its Successful Removal by PROX. *Catalysts* **2017**, *7*, 348. [CrossRef]
18. Guerriero, A.; Peruzzini, M.; Gonsalvi, L. Coordination chemistry of 1,3,5-triaza-7-phosphaadamantane (PTA) and derivatives. Part III. Variations on a theme: Novel architectures, materials and applications. *Coord. Chem. Rev.* **2018**, *355*, 328–361. [CrossRef]
19. Wang, D.; Astruc, D. The Golden Age of Transfer Hydrogenation. *Chem. Rev.* **2015**, *115*, 6621–6686. [CrossRef] [PubMed]

20. Dalebrook, A.F.; Gan, W.; Grasemann, M.; Moret, S.; Laurenczy, G. Hydrogen Storage: Beyond Conventional Methods. *Chem. Commun.* **2013**, *49*, 8735–8751. [CrossRef] [PubMed]
21. Dong, H.; Liu, J.; Ma, L.; Ouyang, L. Chitosan Aerogel Catalyzed Asymmetric Aldol Reaction in Water: Highly Enantioselective Construction of 3-Substituted-3-hydroxy-2-oxindoles. *Catalysts* **2016**, *6*, 186. [CrossRef]
22. Zhang, T.; Chu, S.; Li, J.; Wang, L.; Chen, R.; Shao, Y.; Liu, X.; Ye, M. Efficient Degradation of Aqueous Carbamazepine by Bismuth Oxybromide-Activated Peroxide Oxidation. *Catalysts* **2017**, *7*, 315. [CrossRef]
23. Nakano, S.-I.; Horita, M.; Kobayashi, M.; Sugimoto, N. Catalytic Activities of Ribozymes and DNAzymes in Water and Mixed Aqueous Media. *Catalysts* **2017**, *7*, 355. [CrossRef]
24. Nishiwaki, K.; Kobayashi, M.; Takeuchi, T.; Matuoto, K.; Osakada, K. Nieuwland catalysts: Investigation of structure in the solid state and in solution and performance in the dimerization of acetylene. *J. Mol. Catal. A Chem.* **2001**, *175*, 73–81. [CrossRef]
25. You, Y.; Luo, J.; Xie, J.; Dai, B. Effect of Iminodiacetic Acid-Modified Nieuwland Catalyst on the Acetylene Dimerization Reaction. *Catalysts* **2017**, *7*, 394. [CrossRef]

Article

Chitosan Aerogel Catalyzed Asymmetric Aldol Reaction in Water: Highly Enantioselective Construction of 3-Substituted-3-hydroxy-2-oxindoles

Hui Dong [1,2], Jie Liu [2], Lifang Ma [1,]* and Liang Ouyang [2,]*

[1] School of Chemical Engineering, Sichuan University, Chengdu 610065, China; donghui0553@163.com
[2] State Key Laboratory of Biotherapy, West China Hospital, Sichuan University, and Collaborative Innovation Center for Biotherapy, Chengdu 610041, China; liujie2011@scu.edu.cn
* Correspondence: mlfang11@scu.edu.cn (L.M.); ouyangliang@scu.edu.cn (L.O.);
   Tel.: +86-28-8540-5221 (L.M.); +86-28-8550-3817 (L.O.)

Academic Editor: Luca Gonsalvi
Received: 3 October 2016; Accepted: 21 November 2016; Published: 28 November 2016

Abstract: A chitosan aerogel catalyzed asymmetric aldol reaction of ketones with isatins in the presence of water is described. This protocol was found to be environmentally benign, because it proceeds smoothly in water and the corresponding aldol products were obtained in excellent yields with good enantioselectivities.

Keywords: chitosan aerogel; aldol reaction; water; isatin

1. Introduction

The 3-substituted-3-hydroxy-2-oxindoles have a stereogenic quaternary center at the C-3 position and a core unit that appears in many natural products and biologically active compounds [1–8]. Representative examples are: TMC-95A [9,10], Dioxibrassinine [11,12], SM-130686 [13], 3'- Hydroxygluoisatisin [14], and Convolutamydines (Figure 1) [15]. Consequently, the 3-substituted-3-hydroxy-2-oxindole framework has been an intensively investigatedsynthetic target. To construct 3-substituted-3-hydroxy-2-oxindoles, asymmetric aldol reaction has been considered one of the most powerful and efficient measures for the formation of carbon–carbon bond at C-3 position [16–21]. In this context, the asymmetric aldol reaction between isatin and carbonyl compounds has attracted much attention. As a pioneering work in this field, Tomasini and coworkers demonstrated the enantioselective aldol reaction of isatin with acetone catalyzed by a dipeptide-based organocatalyst [22]. Along these lines, Toru et al. employed sulfonamides as catalysts for the enantioselective aldol reaction of acetaldehyde with isatin, and successfully achieved the first highly enantioselective crossed-aldol reaction of acetaldehydes with ketones [23]. Later on, Zhao et al. described the utilization of quinidine thiourea for the highly enantioselective synthesis of 3-alkyl-3-hydroxyindolin-2-ones [24]. Lin et al. disclosed the enzymatic enantioselective aldol reaction of isatin derivatives with cyclic ketones, which produced products in high yields with moderately good stereoselectivity [25]. Very recently, natural amino acid salts were successfully developed to catalyze direct aldol reactions of isatin with ketones [26]. Despite this reported success, it is still important and desirable to develop new catalysts with operational simplicity and high catalytic efficiency for asymmetric aldol reactions.

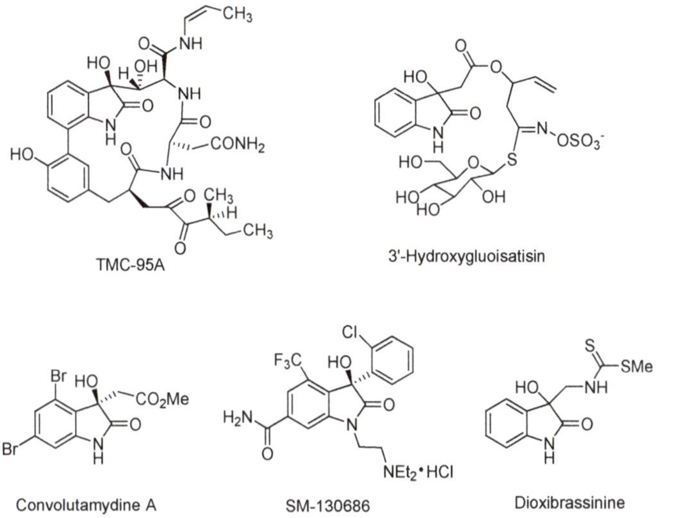

TMC-95A

3'-Hydroxygluoisatisin

Convolutamydine A

SM-130686

Dioxibrassinine

Figure 1. Representative examples of 3-substituted-3-hydroxy-2-oxindoles.

Recently, considerable focus has been placed on environmentally-friendly and sustainable resources and processes. In this regard, natural materials have been used directly as supports for catalytic applications, which has made this approach a very attractive strategy. In particular, biopolymers are a diverse and versatile class of materials that are inexpensive and abundant in nature [27]. Chitosan is a very abundant biopolymer obtained from the alkaline deacetylation of chitin, which is ubiquitous in the exoskeletons of crustaceans, the cuticles of insects and the cell walls of most fungi [28]. Chitosan functionalization is based on the presence of amino groups, which easily react with electrophilic reagents such as aldehydes, acid chlorides, acid anhydrides and epoxides [29–32]. Chitosan is a chiral polyamine and exhibits good flexibility, insolubility in many solvents and an inherent chirality and affinity for metal ions [33–36]. On the other hand, chitosan is an excellent candidate for building heterogeneous catalysts, since it can act as a support for chiral organic frameworks [37,38]. In addition, there are various advantages to using chitosan to catalyze reactions in water, which is a universally environmentally-friendly solvent.

Although chitosan possesses these properties, the direct use of chitosan in base catalysis has been very poorly investigated. In 2006, Kantam et al. reported the use of chitosan hydrogels as a green and recyclable catalyst for the aldol and Knoevenagel reactions [39]. Since then, as an ideal alternative to organocatalysts, chitosan aerogels were utilized to catalyze the asymmetric aldol reaction in water, giving the desired products in high yields with good stereoselectivity and recyclability [40]. Not long ago, chitosan-supported cinchonine was developed as an organocatalyst for the direct asymmetric aldol reaction in water and this catalyst could be easily recovered and reused several times without a significant loss in activity [41]. In continuation of our previous efforts on asymmetric direct aldol reaction of isatins with ketones [26], herein, we report on the synthesis of 3-substituted-3-hydroxy-2-oxindoles catalyzed by chitosan aerogel in the presence of water.

2. Results and Discussion

The direct asymmetric aldol reaction of isatin and hydroxyacetone was selected as a template reaction to optimize the conditions for the reaction catalyzed using the chitosan aerogel. Using the optimized conditions, the desired product was obtained in high yield with excellent stereoselectvity and relative configurations assigned by comparison with previously reports [42,43].

Initially, the aldol reaction with hydroxyacetone **2a** was examined as the donor substrate and isatin **1a** as the acceptor using 10 mol % catalyst at room temperature and the results of this reaction are shown in Table 1.

Table 1. Screening of the solvents for the enantioselective aldol reaction of isatin and hydroxyacetone catalyzed by chitosan aerogel [a].

Entry	Solvent	Time (h)	Yield (%) [b]	syn:anti [c]	ee (%) [d]
1	EtOAc	3	92	71:29	11 (16)
2	MeCN	2	90	72:28	21 (20)
3	Et$_2$O	3	94	52:48	7 (4)
4	CHCl$_3$	2	92	65:35	31 (30)
5	PhMe	3	95	52:48	35 (22)
6	DCM	2	97	65:35	34 (10)
7	Hexane	1.5	93	65:35	40 (20)
8	H$_2$O	2	90	73:27	37 (57)
9	EtOH	2	80	70:30	17 (20)
10	MeOH	2	88	67:33	29 (20)
11	Dioxane	7	70	61:39	20 (13)
12	*i*-PrOH	6	70	57:43	18 (5)

[a] Unless otherwise indicated, all reactions were carried out with isatin **1a** (0.5 mmol), hydroxyacetone **2a** (2.5 mmol), and 0.05 mmol chitosan aerogel (10 mol %) at room temperature for the specified time; [b] Isolated yields after purification by flash column; [c] Determined by chiral HPLC analysis or ^{1}H NMR analysis of the crude mixture; [d] Determined by chiral HPLC analysis (results in parentheses refer to the minor diastereoisomer).

The data in Table 1 showed that the catalytic activity and stereoselectivity of the reaction were influenced by the reaction media. The solvents employed included EtOAc, MeCN and Et$_2$O which produced the product **3a** with 90%–94% yields, but with poor enantiomeric excess (ee) (Table 1, entries 1–3). Other solvents that were also tested produced **3a** in excellent yields (90%–97%) and with good ee (31%–40%) (Table 1, entries 4–8), but also produced high diastereoselectivity (syn:ant = 73:27) with water as the solvent (Table 1, entry 8). Although physical properties between hexane and water are greatly different, the product **3a** in yield and enantioselectivity in hexane were similar to that in water, this might be because the chitosan aerogel exhibited higher catalytic activity in water and hexane. In addition, compared to water, the solvents MeOH and EtOH exhibited lower yields (88% and 80%) and enantioselectivity (29% ee and 17% ee) (Table 1, entries 9–10). In addition, when dioxane and *i*-PrOH were used as the solvents, the results were unsatisfactory. Therefore, water was chosen as the solvent due to its good performance in this reaction.

After screening the solvents, the effect of the donor (hydroxyacetone) on the model aldol reaction was investigated. As summarized in Table 2, increasing the amount of hydroxyacetone from 5 to 20 equivalents led to a significant increase in the yields (88% up to 99%) and enantioselectivity (33% ee up to 44%) (Table 2, entries 1–3). A further increase in the hydroxyacetone dosage to over 20 equivalents resulted in a decrease in the enantioselectivity, albeit with excellent yields (99%) (Table 2, entry 4). This suggested that the donor dosage within a certain range can improve the chemoselectivity. Accordingly, 20 equivalents of hydroxyacetone was used as the optimum dosage.

Table 2. Effect of the amount of isatin for the enantioselective aldol reaction in the presence of water [a].

chitosan aerogel
H_2O, rt

1a   2a   3a

Entry	Hydroxyacetone (Equivalents)	Time (h)	Yield (%) [b]	syn:anti [c]	ee (%) [d]
1	5	2.5	88	71:29	33 (43)
2	10	2	96	72:28	40 (51)
3	20	2	99	57:43	44 (59)
4	40	1.5	99	53:47	31 (30)

[a] Unless otherwise indicated, all reactions were carried out with isatin **1a** (0.5 mmol), hydroxyacetone **2a** and 0.05 mmol chitosan aerogel (10 mol %) in water (0.5 mL) stirred at room temperature for the specified time; [b] Isolated yields after purification by flash column; [c] Determined by chiral HPLC analysis or ^{1}H NMR analysis of the crude mixture; [d] Determined by chiral HPLC analysis (results in parentheses refer to the minor diastereoisomer).

To further enhance the yield and enantioselectivity of **3a**, the effect of different types of additives was investigated (Table 3). The results showed that 2,5-dihydroxybenzoic acid was the most effective additive, producing product **3a** in a 96% yield with 66% ee (Table 3, entry 16). By contrast, compared to 2,5-dihydroxybenzoic acid, some additives such as sulfamic acid, formic acid and 1,1′-bi-2-naphthol exhibited higher yields but lower enantioselectivity (Table 3, entries 1, 2 and 15). It is also conceivable that the 2,5-dihydroxybenzoic acid that exhibited higher enantioselectivity might also provide an additional asset for chitosan, favoring the recognition of this reagent by the catalyst. Other additives were also tested using the template reaction, which produced product **3a** in good yields along with lower enantioselectivity (Table 3, entries 3–14). Moreover, considering that the reaction temperature is related to the enantioselectivity, the reaction was conducted at 0 °C. Fortunately, it was found that the enantioselectivity of the product was significantly increased by maintaining the reaction temperature at 0 °C (Table 3, entry 17). Hence, the optimum conditions for this reaction were found to be the use of 2,5-dihydroxybenzoic acid as an additive and maintaining the reaction at 0 °C.

Using these optimized conditions, the scope of this reaction was studied and the results are summarized in Table 4.The results showed that the isatin and hydroxyacetone gave the corresponding aldol product **3a** in high yield and good ee (Table 4, **3a**). Isatin containing weaker electron-withdrawing groups such as halogens, also gave excellent yields, but lower enantioselectivity (Table 4, **3b** and **3e**). Unfortunately, although the product **3c** was obtained inexcellent yield, the ee could not be determined. Interestingly, isatin containing strong electron-withdrawing substituents, such as 5-nitroisatin, reacted easily with hydroxyacetone to give **3d** in high yield (92%) and good enantioselectivity (92% ee), along with excellent diastereoselectivity (Table 4, **3d**). Subsequently, *N*-benzylisatins with various substitution patterns were studied, and the corresponding products **3f**–**3i** were obtained in better yields (93%–97%) and good enantioselectivity (72%–94% ee) (Table 4, **3f**–**3i**). However, the product **3j** showed lower enantioselectivity (Table 4, **3j**). *N*-methylisatin was also employed to produce product **3k** in higher yield and ee (Table 4, **3k**). When *N*-boc and *N*-acethylisatins were used, the corresponding products **3l** and **3m** exhibiting an enantiomeric excess could not be clearly identified. Finally, using methoxyacetone as the donor substrate, the resulting products **3n**–**3p** were obtained in excellent yields and enantioselectivity (Table 4, **3n**–**3p**).

Table 3. Effects of additives for the enantioselective aldol reaction in the presence of water [a].

Entry	Additive	Time (h)	Yield (%) [b]	syn:anti [c]	ee (%) [d]
1	sulfamic acid	2.5	98	56:44	2 (1)
2	formic acid	4	97	68:32	38 (47)
3	*p*-toluenesulfonic acid	2.5	93	60:40	11 (20)
4	4 Å molecular sieves(10 mg) [e]	2.5	90	54:46	11 (26)
5	4 Å molecular sieves(20 mg) [e]	2.5	93	61:39	10 (12)
6	4 Å molecular sieves(30 mg) [e]	4.5	96	72:28	31 (56)
7	acetic acid glacial	2.5	94	59:41	17 (22)
8	L-proline	2.5	95	64:36	24 (29)
9	benzoic acid	2.5	97	57:43	16 (18)
10	stearic acid	5	94	51:49	11 (4)
11	3-nitrobenzoic acid	4	92	61:39	29 (25)
12	2,4-dinitrophenol	4	94	69:31	47 (60)
13	polyethylene glycol	6	87	65:35	34 (41)
14	oxalic acid	2.5	82	58:42	46 (50)
15	1,1′-bi-2-naphthol	3	98	68:32	63 (62)
16	2,5-dihydroxybenzoic acid	2.5	96	69:31	66 (72)
17 [f]	2,5-dihydroxybenzoic acid	48	90	53:47	75 (80)

[a] Unless otherwise indicated, all reactions were carried out with isatin **1a** (0.5 mmol), hydroxyacetone **2a** (10 mmol), 0.05 mmol chitosan aerogel (10 mol %) and 0.05 mmol additive (10 mol %) in water (0.5 mL) stirred at room temperature for the specified time; [b] Isolated yields after purification by flash column; [c] Determined by chiral HPLC analysis or ^{1}H NMR analysis of the crude mixture; [d] Determined by chiral HPLC analysis(results in parentheses refer to the minor diastereoisomer); [e] Pulverized without activation; [f] The reaction was conducted at 0 °C.

Table 4. Asymmetric aldol reaction between various isatins and ketones catalyzed by chitosan aerogel under optimized conditions [a,b,c,d].

3a

90%, 53:47, 75%ee

3b

90%, 83:17, 36%ee

3c

93%, 71:29

3d

92%, >99:1, 92%ee

Table 4. *Cont.*

1a-m	**2a-b**	**3a-p**

3e	3f	3g	3h
95%, 53:47, 66%ee	93%, 53:47, 72%ee	96%, 62:38, 74%ee	94%, 90:10, 73%ee

3i	3j	3k	3l
97%, 96:4, 94%ee	91%, 93:7, 34%ee	98%, 52:48, 77%ee	90%, 57:43

3m	3n	3o	3p
91%, 77:23	91%, 81:19, 87%ee	92%, 88:12, 94%ee	93%, 75:25, 87%ee

[a] All reactions were carried out with isatin **1a** (0.5 mmol), hydroxyacetone **2a** (10 mmol), 0.05 mmol chitosan aerogel (10 mol %) and 0.05 mmol 2,5-dihydroxybenzoic acid (10 mol %) in water (0.5 mL) stirred at 0 °C for 48 h; [b] Isolated yields after purification by flash column; [c] Determined by chiral HPLC analysis or ^{1}H NMR analysis of the crude mixture; [d] Determined by chiral HPLC analysis.

3. Materials and Methods

3.1. General Methods

All solvents and reagents in this work were acquired from different commercial sources and used without further purification. Chitosan aerogel microspheres were prepared as in previous literature [32]. Thin layer chromatography (TLC) was conducted on GF254 silica gel plates, which were visualized by UV at 254 nm. Column chromatography separations were performed using silica gel 300–400 mesh. Chiral High-performance liquid chromatography (HPLC) analysis was conducted with a Waters Alliance 2695 instrument (Waters corporation, Milford, MA, USA), using a UV–visible light (Vis) Waters PDA 2998 detector (Waters corporation), and working at 254 nm. ^{1}H NMR spectra were recorded on a Bruker AM400 NMR spectrometer (Bruker corporation, Karlsruhe, Germany), and NMR spectra were obtained as CDCl$_3$ solutions (reported in ppm), using chloroform as the reference standard (7.26 ppm) or dimethyl sulfoxide-d_6 (DMSO-d_6) (2.50 ppm). High-resolution mass spectrometry (HRMS) data were recorded using a Waters Q-Tof premier mass spectrometer (Waters corporation).

3.2. General Procedure for the Asymmetric Aldol Reaction of Isatins with Ketones

A reaction mixture of isatin (0.5 mmol), ketone (10 mmol), chitosan aerogel beads (10 mol %) and 2,5-dihydroxybenzoic acid (10 mol %) in water (0.5 mL) was stirred at 0 °C until the complete conversion of the starting material. Then the solvent was removed in vacuo to give the crude product and purified by column chromatography on silica gel (petroleum ether/ethyl acetate) or crystallization from petroleum ether/ethyl acetate to afford the desired compounds.

4. Conclusions

In conclusion, an environmentally-friendly enantioselective aldol reaction to construct the 3-substituted-3-hydroxy-2-oxindoles is established by using isatins and ketones as starting materials. In this reaction, chitosan aerogel is successfully employed as a green organocatalyst and works smoothly in the presence of water. The reaction has a large substrate scope and the corresponding products are all produced in high yields and with high chemoselectivity. Moreover, 2,5-dihydroxybenzoic acid was found to be an effective additive to modulate the asymmetric aldol reactions. Further studies of this system can broaden the scope of this reaction to other ketone donors.

Supplementary Materials: The following are available online at www.mdpi.com/2073-4344/6/12/186/s1, Figures S1–S16: ^{1}H NMR and ^{13}C NMR analysis of compound **3a–3p**, Figures S17–S29: Chiral High-performance liquid chromatography (HPLC) analysis of compound **3a–3p**.

Acknowledgments: This work was supported by the National Natural Science Foundation of China (81473091, 81673290) and the Fundamental Research Funds for the Central Universities and Distinguished Young Scholars of Sichuan University (2015SCU04A41).

Author Contributions: Lifang Ma and Liang Ouyang conceived and designed the experiments; Hui Dong performed the experiments; Jie Liu analyzed the data; Liang Ouyang and Jie Liu contributed reagents/materials/analysis tools; Hui Dong and Liang Ouyang wrote the paper. All authors read and approved the final manuscript.

Conflicts of Interest: The authors declare no conflict of interest.

References

1. Rasmussen, H.B.; Macleod, J.K. Total synthesis of donaxaridine. *J. Nat. Prod.* **1997**, *60*, 1152–1154. [CrossRef]
2. Hibino, S.; Choshi, T. Simple indole alkaloids and those with a nonrearranged monoterpenoid unit. *Nat. Prod. Rep.* **2001**, *18*, 66–87. [CrossRef] [PubMed]
3. Tang, Y.Q.; Sattler, I.; Thiericke, R.; Grabley, S.; Feng, X.Z. Maremycins C and D, new diketopiperazines, and maremycins E and F, novel polycyclic *spiro*-indole metabolites isolated from *Streptomyces* sp. *Eur.J. Org. Chem.* **2001**, 261–267. [CrossRef]
4. Hewawasam, P.; Erway, M.; Moon, S.L.; Knipe, J.; Weiner, H.; Boissard, C.G.; Post-Munson, D.J.; Gao, Q.; Huang, S.; Gribkoff, V.K.; et al. Synthesis and structure−activity relationships of 3-aryloxindoles: A new class of calcium-dependent, large conductance potassium (Maxi-K) channel openers with neuroprotective properties. *J. Med. Chem.* **2002**, *45*, 1487–1499. [CrossRef] [PubMed]
5. Nicolaou, K.C.; Rao, P.B.; Hao, J.; Reddy, M.V.; Rassias, G.; Huang, X. The second total synthesis of diazonamide A. *Angew. Chem. Int. Ed.* **2003**, *42*, 1753–1758. [CrossRef] [PubMed]
6. Suzuki, H.; Morita, H.; Shiro, M.; Kobayashi, J. Celogentin K, a new cyclic peptide from the seeds of Celosia argentea and X-ray structure of moroidin. *Tetrahedron* **2004**, *60*, 2489–2495. [CrossRef]

7. Chen, W.B.; Du, X.L.; Cun, L.F.; Zhang, X.M.; Yuan, W.C. Highly enantioselective aldol reaction of acetaldehyde and isatins only with 4-hydroxydiarylprolinol as catalyst: Concise stereoselective synthesis of (*R*)-convolutamydines B and E, (−)-donaxaridine and (*R*)-chimonamidine. *Tetrahedron* **2010**, *66*, 1441–1446. [CrossRef]

8. Niu, R.; Xiao, J.; Liang, T.; Li, X.W. Facile synthesis of azaarene-substituted 3-hydroxy-2-oxindoles via brønsted acid catalyzed sp^3 C–H functionalization. *Org. Lett.* **2012**, *14*, 676–679. [CrossRef] [PubMed]

9. Koguchi, Y.; Kohno, J.; Nishio, M.; Takahashi, K.; Okuda, T.; Ohnuki, T.; Komatsubara, S. TMC-95A, B, C, and D, novel proteasome inhibitors produced by *Apiospora montagnei* Sacc. TC 1093 taxonomy, production, isolation, and biological activities. *J. Antibiot.* **2000**, *53*, 105–109. [CrossRef] [PubMed]

10. Kohno, J.; Koguchi, Y.; Nishio, M.; Nakao, K.; Kuroda, M.; Shimizu, R.; Ohnuki, T.; Komatsubara, S. Structures of TMC-95A−D: Novel proteasome inhibitors from *Apiospora montagnei* Sacc. TC 1093. *J. Org. Chem.* **2000**, *65*, 990–995. [CrossRef] [PubMed]

11. Monde, K.; Sasaki, K.; Shirata, A.; Takasugi, M. Brassicanal C and two dioxindoles from cabbage. *Phytochemistry* **1991**, *30*, 2915–2917. [CrossRef]

12. Suchý, M.; Kutschy, P.; Monde, K.; Goto, H.; Harada, N.; Takasugi, M.; Dzurilla, M.; Balentova, E. Synthesis, absolute configuration, and enantiomeric enrichment of a cruciferous oxindolephytoalexin, (*S*)-(−)-spirobrassinin, and its oxazoline analog. *J.Org. Chem.* **2001**, *66*, 3940–3947. [CrossRef] [PubMed]

13. Tokunaga, T.; Hume, W.E.; Umezome, T.; Okazaki, K.; Ueki, Y.; Kumagai, K.; Hourai, S.; Nagamine, J.; Seki, H.; Taiji, M.; et al. Oxindole derivatives as orally active potent growth hormone secretagogues. *J. Med. Chem.* **2001**, *44*, 4641–4649. [CrossRef] [PubMed]

14. Fréchard, A.; Fabre, N.; Péan, C.; Montaut, S.; Fauvel, M.T.; Rollin, P.; Fourasté, I. Novel indole-type glucosinolates from woad (*Isatis tinctoria* L.). *Tetrahedron Lett.* **2001**, *42*, 9015–9017. [CrossRef]

15. Kamano, Y.; Zhang, H.P.; Ichihara, Y.; Kizu, H.; Komiyama, K.; Pettit, G.R. Convolutamydine A, a novel bioactive hydroxyoxindole alkaloid from marine bryozoan *Amathia convolute. Tetrahedron Lett.* **1995**, *36*, 2783–2784. [CrossRef]

16. Garden, S.J.; Silva, R.B.; Pinto, A.C. A versatile synthetic methodology for the synthesis of tryptophols. *Tetrahedron* **2002**, *58*, 8399–8412. [CrossRef]

17. Chen, J.R.; Liu, X.P.; Zhu, X.Y.; Li, L.; Qiao, Y.F.; Zhang, J.M.; Xiao, W.J. Organocatalytic asymmetric aldol reaction of ketones with isatins: Straightforward stereoselective synthesis of 3-alkyl-3-hydroxyindolin-2-ones. *Tetrahedron* **2007**, *63*, 10437–10444. [CrossRef]

18. Chen, W.B.; Liao, Y.H.; Du, X.L.; Zhang, X.M.; Yuan, W.C. Catalyst-free aldol condensation of ketones and isatins under mild reaction conditions in DMF with molecular sieves 4 Å as additive. *Green Chem.* **2009**, *11*, 1465–1476. [CrossRef]

19. Zhu, B.; Zhang, W.; Lee, R.; Han, Z.; Yang, W.; Tan, D.; Huang, K.W.; Jiang, Z. Direct asymmetric vinylogousaldol reaction of allyl ketones with isatins: divergent synthesis of 3-hydroxy-2-oxindole derivatives. *Angew. Chem. Int. Ed.* **2013**, *52*, 6666–6670. [CrossRef] [PubMed]

20. Kumar, A.S.; Ramesh, P.; Kumar, G.S.; Swetha, A.; Nanubolu, J.B.; Meshram, H.M. A Ru(III)−catalyzed α-cross-coupling aldol type addition reaction of activated olefins with isatins. *RSC Adv.* **2016**, *6*, 1705–1709. [CrossRef]

21. Chen, Q.; Tang, Y.; Huang, T.; Liu, X.; Lin, L.; Feng, X.M. Copper/Guanidine-catalyzed asymmetric alkynylation of isatins. *Angew. Chem. Int. Ed.* **2016**, *55*, 5286–5289. [CrossRef] [PubMed]

22. Luppi, G.; Cozzi, P.G.; Monari, M.; Kaptein, B.; Broxterman, Q.B.; Tomasini, C. Dipeptide-catalyzed asymmetric aldol condensation of acetone with (*N*-alkylated) isatins. *J. Org. Chem.* **2005**, *70*, 7418–7421. [CrossRef] [PubMed]

23. Nakamura, S.; Hara, N.; Nakashima, H.; Kubo, K.; Shibata, N.; Toru, T. First enantioselective synthesis of (*R*)-convolutamydine B and E with *N*-(heteroarenesulfonyl) prolinamides. *Chem. Eur. J.* **2009**, *15*, 6790–6793.

24. Guo, Q.; Bhanushali, M.; Zhao, C. Quinidine thiourea-catalyzed aldol reaction of unactivated ketones: Highly enantioselective synthesis of 3-alkyl-3-hydroxyindolin-2-ones. *Angew. Chem. Int. Ed.* **2010**, *49*, 9460–9464. [CrossRef] [PubMed]

25. Liu, Z.Q.; Xiang, Z.W.; Shen, Z.; Wu, Q.; Lin, X.F. Enzymatic enantioselective aldol reactions of isatin derivatives with cyclic ketones under solvent-free conditions. *Biochimie* **2014**, *101*, 156–160. [CrossRef] [PubMed]

26. Chen, G.; Ju, Y.; Yang, T.; Li, Z.; Wei, A.; Sang, Z.; Liu, J.; Lou, Y. Natural amino acid salt catalyzed aldol reactions of isatins with ketones: Highly enantioselective construction of 3-alkyl-3-hydroxyindolin-2-ones. *Tetrahedron Asymmetry* **2015**, *26*, 943–947. [CrossRef]

27. Kaplan, D.L. *Biopolymers from Renewable Resources*; Springer: Berlin, Germany, 1998.

28. Guibal, E. Heterogeneous catalysis on chitosan-based materials: A review. *Prog. Polym. Sci.* **2005**, *30*, 71–109.

29. Roberts, G.A.F.; Taylo, K.E. Chitosan gels, 3. The formation of gels by reaction of chitosan with glutaraldehyde. *Makromol. Chem.* **1989**, *190*, 951–960. [CrossRef]

30. Wei, Y.C.; Hudson, S.M.; Mayer, J.M.; Kaplan, D.L.J. The crosslinking of chitosan fibers. *Polym. Sci. Part A* **1992**, *30*, 2187–2193. [CrossRef]

31. Zeng, X.; Ruckenstein, E. Trypsin purification by *p*-aminobenzamidine immobilized on macroporous chitosan membranes. *Ind. Eng. Chem. Res.* **1998**, *37*, 159–165. [CrossRef]

32. Quignard, F.; Valentin, R.; Renzo, F.D. Aerogel materials from marine polysaccharides. *New J. Chem.* **2008**, *32*, 1300–1310. [CrossRef]

33. Quignard, F.; Choplin, A.; Domard, A. Chitosan: A natural polymeric support of catalysts for the synthesis of fine chemicals. *Langmuir* **2000**, *16*, 9106–9108. [CrossRef]

34. Sun, W.; Xia, C.G.; Wang, H.W. Efficient heterogeneous catalysts for the cyclopropanation of olefins. *New J. Chem.* **2002**, *26*, 755–758. [CrossRef]

35. Hardy, J.J.E.; Hubert, S.; Macquarrie, D.J.; Wilson, A.J. Chitosan-based heterogeneous catalysts for Suzuki and Heck reactions. *Green Chem.* **2004**, *6*, 53–56. [CrossRef]

36. Chtchigrovsky, M.; Primo, A.; Gonzalez, P.; Molvinger, K.; Robitzer, M.; Quignard, F.; Taran, F. Functionalized chitosan as a green, recyclable, biopolymer-supported catalyst for the [3 + 2] Huisgen cycloaddition. *Angew. Chem. Int. Ed.* **2009**, *48*, 5916–5920. [CrossRef] [PubMed]

37. Macquarrie, D.J.; Hardy, J.J.E. Applications of functionalized chitosan in catalysis. *Ind. Eng. Chem. Res.* **2005**, *44*, 8499–8520. [CrossRef]

38. Zhang, H.; Zhao, W.; Zou, J.; Yi, L.; Li, R.; Cui, Y. Aldol reaction catalyzed by a hydrophilic catalyst in aqueous micelle as an enzyme mimic system. *Chirality* **2009**, *21*, 492–496. [CrossRef] [PubMed]

39. Reddy, K.R.; Rajgopal, K.; Maheswari, C.U.; Kantam, M.L. Chitosan hydrogel: A green and recyclable biopolymer catalyst for aldol and Knoevenagel reactions. *New J. Chem.* **2006**, *30*, 1549–1552. [CrossRef]

40. Gioia, C.; Ricci, A.; Bernardi, L.; Bourahla, K.; Tanchoux, N.; Robitzer, M.; Quignard, F. Chitosan aerogel beads as a heterogeneous organocatalyst for the asymmetric aldol reaction in the presence of water: An assessment of the effect of additives. *Eur. J. Org. Chem.* **2013**, 588–594. [CrossRef]

41. Zhao, W.; Qu, C.; Yang, L.; Cui, Y. Chitosan-supported cinchonine as an efficient organocatalyst for direct asymmetric aldol reaction in water. *Chin. J. Catal.* **2015**, *36*, 367–371. [CrossRef]

42. Ricci, A.; Bernardi, L.; Gioia, C.; Vierucci, S.; Robitzer, M.; Quignard, F. Chitosan aerogel: A recyclable, heterogeneous organocatalyst for the asymmetric direct aldol reaction in water. *Chem. Commun.* **2010**, *46*, 6288–6290. [CrossRef] [PubMed]

43. Tanimura, Y.; Yasunaga, K.; Ishimaru, K. Asymmetric aldol reaction using a very simple primary amine catalyst: divergent stereoselectivity by using 2,6-difluorophenyl moiety. *Tetrahedron* **2014**, *70*, 2816–2821. [CrossRef]

 *catalysts*

Review

Unconventional Approaches Involving Cyclodextrin-Based, Self-Assembly-Driven Processes for the Conversion of Organic Substrates in Aqueous Biphasic Catalysis

Frédéric Hapiot * and Eric Monflier

Université Artois, CNRS, Centrale Lille, ENSCL, Université Lille, UMR 8181,
Unité de Catalyse et de Chimie du Solide (UCCS), F-62300 Lens, France; eric.monflier@univ-artois.fr
* Correspondence: frederic.hapiot@univ-artois.fr; Tel.: +33-(0)-321791773

Academic Editor: Luca Gonsalvi
Received: 2 May 2017; Accepted: 26 May 2017; Published: 2 June 2017

Abstract: Aqueous biphasic catalysis is a convenient approach to convert organic, partially soluble molecules in water. However, converting more hydrophobic substrates is much more challenging as their solubility in water is extremely low. During the past ten years, substantial progress has been made towards improving the contact between hydrophobic substrates and a hydrophilic transition-metal catalyst. The main cutting-edge approaches developed in the field by using cyclodextrins as a supramolecular tool will be discussed and compared in this short review.

Keywords: catalysis; cyclodextrins; supramolecular chemistry

1. Introduction

The 21st century has seen the emergence of new organometallic catalytic systems involving multiple components held together by weak interactions (hydrogen bonding, metal–ligand, π–π stacking, electrostatic and Van der Waals interactions, and hydrophobic and solvatophobic effects). A subdiscipline of catalysis thus emerged aiming at tackling the higher complexity of catalyst design using the tools of supramolecular chemistry. This new disciplinary field is referred to as "supramolecular catalysis" [1,2]. The two key concepts underlying this discipline are molecular recognition and self-organization. Molecular recognition deals with the mutual affinity of two or more components. Enzymes proceed via molecular recognition as they fulfill multiple functions at the same time through a network of non-covalent interactions spanning from hydrogen bonds to much weaker Van der Waals forces. Additionally, they surround the substrate to force it to react in the desired manner in an organic pocket where hydrophobic forces are exerted. The term self-organization refers to a process in which the internal organization of a system, usually a non-equilibrium system, increases automatically without being directed by an external source. The spontaneous emergence of a spatial structure results from the interactions at work between elements of the system under consideration. The global properties of the obtained system are often very different from those of the individual elements.

From molecular recognition and self-organization, novel catalytic supramolecular systems have emerged. For example, the use of "host" receptors capable of supramolecularly recognizing guest molecules (substrates and/or catalyst) within their cavity showed that benefits could be gained in terms of catalyst implementation and catalytic performances (turnover, chemo- and stereoselectivities, product inhibition, etc.) [3]. Concurrently, weak interactions also proved to be effective to access novel self-assembled catalysts able to adjust their three-dimensional structure to provide good complementarity to the transition state, thus contributing to the chemo-, regio-, and stereoselectivity of the reaction. Most of the strategies developed so far in supramolecular catalysis have been implemented in organic solvents [4,5]. However, aqueous catalytic systems have also been developed. This short review is focused on cyclodextrin (CD)-based, self-assembly-driven catalytic processes taking place in water in the presence of organometallic catalysts. Emphasis is on CD-based self-assembled organometallic catalysts, micellar aggregates, and host–guest catalysis developed during the past 10 years. Through chosen examples, the role of CDs as second-sphere ligands of organometallic catalysts (and first-sphere ligands if applicable) will be discussed. Catalytic processes for which CDs are exclusively considered as first sphere ligands are not discussed herein [6,7]. The cited references are not intended to be an exhaustive list of all the works on the topic but are portals to other publications.

2. CD-Based Organometallic Catalysts Self-Assembled via Hydrophobic Effects

Countless novel supramolecular catalysts have been elaborated using the tools of supramolecular chemistry. Compared to the standard covalent approach, the synthesis of such catalysts was straightforward as the components self-assembled in situ during the catalytic process. Figure 1 depicts the most effective strategies developed to access self-assembled catalysts.

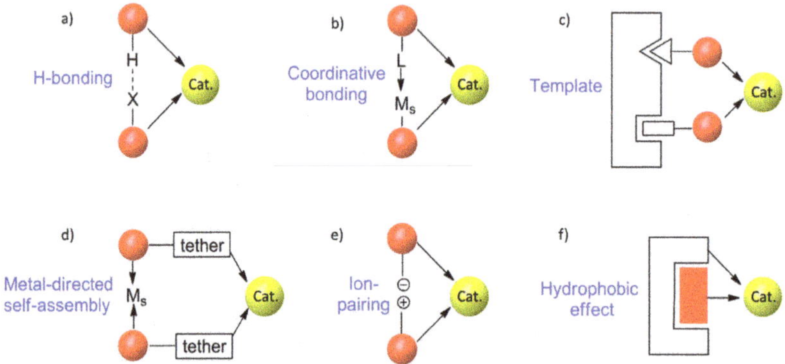

Figure 1. Strategies to elaborate supramolecular catalysts (M_S: structural metal).

Breit introduced one of the first strategies based on hydrogen bonding (Figure 1a) [8]. Concurrently, Reek and van Leeuwen developed supramolecular assemblies through coordination of bidentate ligands (Figure 1c) [9,10]. Takacs elaborated metal-directed self-assemblies (Figure 1d) [11] while Ooi developed the synthesis of ion-paired chiral ligands for organometallic catalysis (Figure 1e) [12]. In aqueous media, one of the most relevant strategy relies on hydrophobic effects (Figure 1f) [13,14]. Our group used CDs as platforms to elaborate new water-soluble bidentate ligands (Figure 2). Innovative water-soluble catalysts are obtained upon inclusion of water-soluble phosphanes in the cavity of modified β-CDs. First described in 2007 for platinum-complexes hydrogenation of water-soluble alkenes [15], the concept of supramolecular CD-based bidentate ligands was extended to the synthesis of rhodium-complexes using mono-*N,N*-dialkylamino-β-CD derivatives and sulfonated phosphanes having a strong affinity for the CD cavity (K_a = 40,000 M^{-1} at 25 °C) [16]. Upon inclusion, the resulting supramolecular phosphorus-nitrogen (PN) bidentate acted as a first-sphere ligand via coordination of both a phosphane and an amine onto the metal and as a second-sphere ligand due to the inclusion of a phosphane inside the CD cavity. In hydroformylation (HF) of styrene under biphasic conditions, such Rh–PN complexes mainly yielded branched aldehydes (95% conversion within 2 h, branched/linear ratio up to 91/9) [17]. CD-based bidentates also acted as first- and second-sphere ligands in aqueous hydrogenation of unsaturated and allylic derivatives [18], and in a domino reaction implying nitrobenzene derivatives subjected consecutively to Pt-catalyzed reduction, Paal–Knorr cyclization and Pt-catalyzed hydrogenation [19].

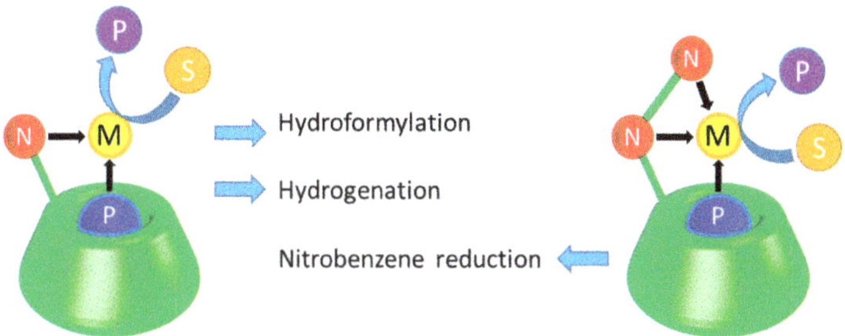

Figure 2. Supramolecular cyclodextrin (CD)-based phosphorus-nitrogen (PN) and phosphorus-nitrogen-nitrogen (PNN) ligands for aqueous catalysis. M: metal; S: substrate; P: Product.

Concurrently to supramolecular systems acting as second- (and possibly first-) sphere ligands, self-assembled systems where CDs only acted as a second-sphere ligand were also developed. For example, CD dimers acted as molecular platforms capable of supramolecularly interacting with both catalyst and substrate through multiple recognition in a confined environment [20]. Compared to monotopic β-CD, the closeness of the involved entities in the surrounding of the dipotic β-CD platform greatly enhanced both the conversion and the chemoselectivity in Rh-catalyzed HF of 1-decene. CDs also helped to convert hydrophobic substrates comprising from 8 to 12 carbons. However, the CD efficacy is significantly reduced when substrates comprising more than 12 carbons are considered. In collaboration with Gonsalvi, Peruzzini and coworkers, our group also developed new phosphanes having both surface activity and ability to coordinate to metals [21]. 1,3,5-triaza-7-phosphaadamantane (PTA) ligand was *N*-substituted by a chemical motive (*para-tert*-butylbenzyl) capable of interacting with the cavity of randomly methylated β-CD (RAME-β-CD) (Figure 3). Contrary to the above supramolecular CD-based PN and phosphorus-nitrogen-nitrogen (PNN) ligands, the supramolecular interaction between the CD host and the PTA ligand should not be too high (K_a = 5440 M^{-1} at 25 °C) as both components should separate easily when increasing the temperature. Depending on the temperature, the amphiphilicity of the phosphane was exalted or reduced upon addition of RAME-β-CD in water. Increasing the temperature resulted in lower CD–phosphane association and subsequent higher concentration of phosphanes in the interfacial layer. For example, in Rh-catalyzed HF of higher olefins, the catalytically active species mostly remained in water at 80 °C due to inclusion of the PTA-ligand within the RAME-β-CD's cavity, thus resulting in a poor catalytic performance. Above 100 °C, the association constant between RAME-β-CD and the PTA ligand greatly decreased, resulting in higher concentrations of "free" species (non-interacting amphiphilic PTA ligand and RAME-β-CD). The availability of the CD cavity was enhanced and the surface activity of the ligand greatly improved, with beneficial effects on the catalytic activity. For example, 98% of 1-decene were converted within 6 h at 120 °C under 50 bar of CO/H$_2$ (substrate/Rh = 500). Additionally, both the aqueous and the organic phases could be simply recovered by cooling down the catalytic system. Indeed, upon cooling, the PTA-ligand accommodated the CD cavity to form a supramolecular complex, which provoked the decantation of both phases.

Adamantyl substituted compounds were also chosen as CD-interacting guests. For example, Leclercq et al. showed that methylated-β-CDs acted as second-sphere ligands of di(1-adamantyl)benzylphosphine to significantly increased the regioselectivity in homogeneous and biphasic HF of allyl alcohol and 1-octene, respectively [22]. The ability of their catalytic system to operate either under homogeneous or biphasic conditions was dictated by the hydrophobic or hydrophilic balance of the catalysts, which itself was defined by the β-CD methylation degree. Another supramolecular complex approach was recently developed which consists in a water-insoluble palladium(II)–dipyrazole complex substituted by an adamantyl moiety [23]. Upon host–guest inclusion of the adamantyl group into the cavity of heptakis(2,6-di-*O*-methyl)-β-CD, high catalytic activity were measured in Suzuki–Miyaura coupling involving hydrophilic aryl bromides and aryl boronic acids in a series of water–organic solvent mixtures. Moreover, using tetrabutyl ammonium bromide as stabilizer, the catalyst was recycled and reused several times. Following a similar approach, a water-soluble glutathione peroxidase mimic was constructed based on the supramolecular host–guest interaction existing between the mono-amino-β-CD and a tellurium compound substituted by two adamantyl groups [24]. The supramolecular system formed hollow vesicle-like aggregates in water, which exhibited enzyme-like catalytic activity and specific recognition ability in the reduction of cumene hydroperoxide by 3-carboxyl-4-nitrobenzenethiol.

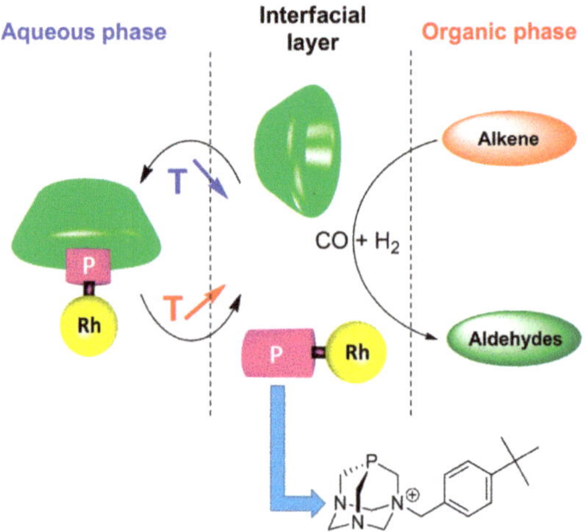

Figure 3. Thermocontrolled catalysis using randomly methylated β-CD (RAME-β-CD) and appropriate phosphane.

3. Control of Micelle Dynamics

Micelles are surfactant-based nano-aggregates acting as nanoreactors by virtue of the confinement of substrates and/or catalyst inside their self-assembled structure [25]. The catalytic performance was significantly improved as the local concentrations in catalyst and substrates were significantly increased at the micelle interface. For example, CDs and imidazolium surfactants were combined to promote the formation of micelles in thermoregulated Rh-catalyzed HF of alkenes [26]. More precisely, the alkyl chain of surfactants such as [C$_{12}$MIM][X] and [C$_{16}$MIM][X] (X = Br or TfO) included into the α-CD's cavity at room temperature while both components dissociated at 80 °C to trigger self-aggregation of imidazolium salts into micelles. In Rh-catalyzed HF of allyl aromatics or styrene, such micelles hosted the substrate into their core while the Rh concentration increased at the micelle surface upon interaction of the imidazolium with the rhodium catalyst, resulting in enhanced conversions (100% conversion at 80 °C under 50 bar of CO/H$_2$). Upon cooling down the solution once the reaction was complete, the inclusion of the imidazolium alkyl chain into the α-CD's cavity provoked the destabilization of the micelles. Through temperature control, the catalytic performance was improved and the rhodium catalyst was recovered.

Grafting dodecyl-imidazolium onto β-CDs was also an effective strategy to access supramolecular hosts capable of self-assembling in neat water [27]. The catalytic properties of the resulting micelles were assessed in Suzuki–Miyaura reaction implying aryl halides and boronic acid derivatives. Interestingly, the dodecyl-imidazolium CD-based micelles allowed converting sterically hindered substrates having strong hydrophobic character into hetero coupling products. Additionally, the system was recycled more than 10 times without significant loss of catalytic activity.

Our group developed the synthesis of amphiphilic phosphanes featuring both surface-activity and ability to coordinate to metal precursors. To overcome the formation of stable emulsions resulting from mixtures of amphiphilic molecules, water and organic substrate, we made use of modified β-CDs. We especially found that addition of stoichiometric proportions of ionic β-CD into amphiphilic phosphane-containing micellar solutions unexpectedly improved the performances of Rh-catalyzed HF of 1-decene while addition of neutral β-CDs resulted in the destructuring of micelles in the post-micellar concentration range [28]. Using ionic β-CDs, an average fourfold rise in catalytic activity was measured (65% conversion of 1-decene within 6 h under 50 bar of CO/H_2 at 80 °C, substrate/Rh = 260) without any influence on the regioselectivity (linear/branched ratio up to 8.6). In that case, complementary properties of the ligand and the surfactant were expressed in a single material. Additionally, the main advantage of the ionic β-CD-based catalytic system relied on the recovery of the aqueous and organic phases once the reaction was complete, especially if randomly methylated β-CD (RAME-β-CD) was used as additives. Amphiphilic phosphane-based aggregates also acted as substrate reservoir in a palladium-catalyzed Tsuji–Trost reaction [29]. The presence of RAME-β-CD in the aqueous compartment improved dynamics of phosphane-based aggregates by supramolecular means (Figure 4). Exchanges between the hydrophobic aggregate core (which contained the substrate) and the aqueous phase (which contained the Pd-catalyst) were significantly favored. Accordingly, the catalytic performance was boosted in a narrow concentration range. RAME-β-CD acted as fluidifier of the phosphane-based micelles. Note that the catalytic performances are dependent on the strength of the supramolecular interaction between the phosphane and the CD. For K_a values over 200,000 M^{-1}, high turn-over frequencies (TOF) are measured for 1:1 RAME-β-CD/phosphane ratio as the aggregate dynamics are favored. Increasing the RAME-β-CD/phosphane ratio results in the formation of stable RAME-β-CD/phosphane complexes. Micelles are then destructured, resulting in a lower TOF. Conversely, weak K_a values (300–600 M^{-1}) lead to a lower amplitude between the highest and the lowest TOF.

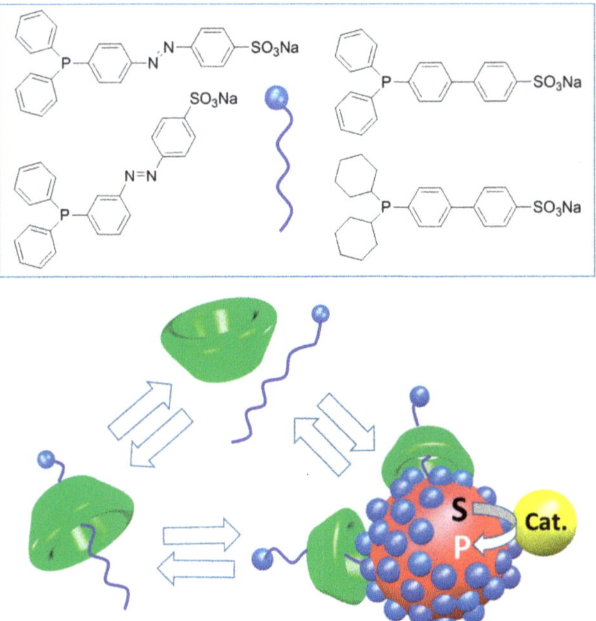

Figure 4. Dynamics of exchange between the hydrophobic micelle core and the aqueous compartment in the presence of RAME-β-CD. S: substrate; P: product; C: organometallic catalyst.

4. Self-Emulsifying Catalytic System

We recently implemented a novel catalytic system involving CD/substrate supramolecular interactions. More precisely, naturally occurring triglycerides were hydroformylated through supramolecular means in the presence of CDs in aqueous medium. During the course of the reaction, a transient supramolecular complex was formed in a well-defined concentration range between triglyceride alkyl chains and CRYSMEB, a methylated β-CD substituted on C2 hydroxyl groups (Figure 5). The triglyceride/water interface was then greatly extended, thus favoring contacts between the triolein carbon–carbon double bond and the rhodium catalyst. In fact, upon inclusion of its alkenyl chain into the CD cavity, the substrate drove its own transformation as the resulting CD/triglyceride supramolecular complexes acted as emulsifiers and helped to convert the triglyceride C=C double bonds in biphasic conditions using a water-soluble organometallic catalyst. At the end of the reaction, the low affinity between the hydroformylated products and CRYSMEB triggered the decantation of products and catalytic solution. The catalytic system was recycled and successfully extended to rhodium-catalyzed HF of commercial oils.

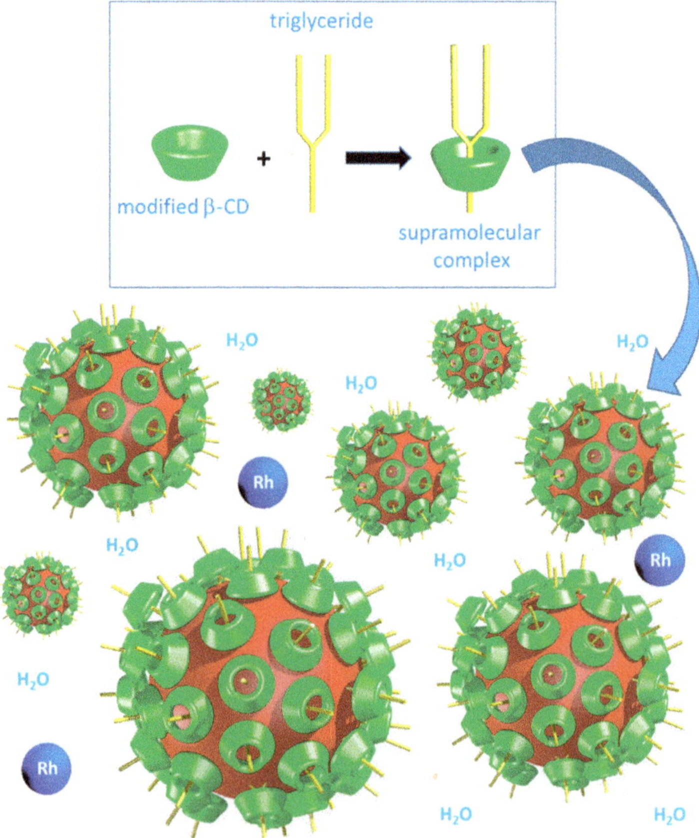

Figure 5. Triglyceride droplets stabilized in water by modified CDs.

4.1. CD-Based Pickering-Like Emulsions

Another approach involving CD-based polymers relies on a thermoresponsive poly(*N*-isopropylacrylamide) (PNIPAM) substituted at its end with RAME-β-CD (Figure 6) [30].

The lower critical solution temperature (LCST) of the resulting modified PNIPAM was 36.6 °C (transition between hydrophilic and hydrophobic state). Its average molecular weight was 16,000 g mol^{-1} (Đ = 1.38) and its hydrodynamic radius was 368 nm at 80 °C. A coil-to-globule transition took place above the LCST, with the formation of micrometer-sized aggregates. In rhodium catalyzed HF of 1-decene and 1-hexadecene, the CD-substituted polymer appeared to be more effective than the separated PNIPAM and RAME-β-CD components. It was shown that RAME-β-CD-based polymer aggregates formed a Pickering-like emulsion (emulsion stabilized by particle) upon adsorption at the aqueous/organic interface. Over time, the aggregates covered the interface to such an extent that the mass transfer was greatly reduced. To overcome the issue, we applied successive heating/cooling cycles to recover the interface fluidity. Cooling down the system below the LCST induced disassembly of the polymer particles. The polymer chains then regained their fluidity and water solubility. The vanishing of the Pickering-like emulsion resulted in the rapid decantation of the biphasic system. Reheating the mixture over the LCST regenerated the Pickering-like emulsion. Complete HF of 1-decene and 1-hexadecene could be implemented through the above step-by-step process.

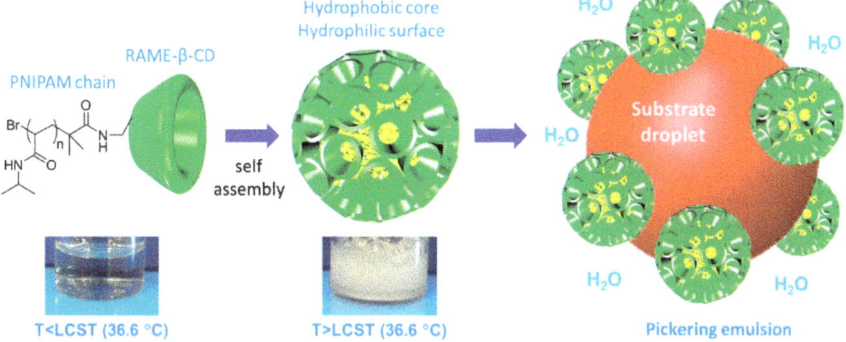

Figure 6. Formation of Pickering emulsion from thermoresponsive poly(*N*-isopropylacrylamide) (PNIPAM) functionalized at the terminal position with RAME-β-CD.

Supramolecular hydrogels consist of three-dimensional networks with non-covalent bonds. They were recently considered as aqueous media likely to contain organometallic catalysts. Interestingly, supramolecular hydrogels consisting of CD-based pseudo-polyrotaxanes resulted from the threading of α-CDs along poly(ethylene glycol) (PEG) chains. Upon threading, α-CDs align to form columnar domains, which subsequently arrange into nanocrystallites. In the presence of an organic phase, the nanocrystallites' adsorption at the water–oil interface resulted in Pickering-like emulsions (Figure 7). For example, the catalytic performance in Rh-catalyzed HF of 1-decene in the sol phase was significantly improved in the presence of α-CD/PEG nanocrystallites (PEG molecular weights of 20,000 g·mol^{-1} (PEG20000) or 35,000 g·mol^{-1} (PEG35000)) [31]. As observed previously, an interface saturation occurred over time. The saturation phenomenon was easily overcome by cooling down the system below the sol–gel temperature and partially depressurizing the autoclave. The Pickering emulsion being broken, dynamics of exchange within the interfacial layer were then greatly improved. Upon temperature and pressure changes, the carbon–carbon double bonds were fully hydroformylated. Addition of RAME-β-CD in well-defined proportion into the hydrogel also substantially improved both the catalytic activity and chemoselectivity [32]. RAME-β-CD also acted as fluidifier of the Pickering-like emulsion and prevented the interface saturation. Dynamics of exchange at the interface were greatly accelerated.

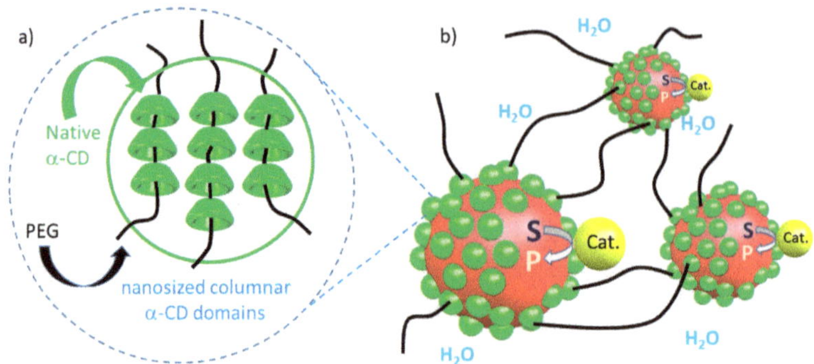

Figure 7. Interfacial catalysis in emulsion using CD/poly(ethylene glycol) (PEG) combination: (**a**) supramolecular hydrogel consisting of nanocrystallites; (**b**) stabilization of substrate droplets in the hydrogel phase via oil-in-water Pickering emulsion. S, substrate; P, product; C, aqueous organometallic catalyst.

4.2. Multivalency with CD Dimers and CD Polymers

Multivalency is a powerful self-assembly pathway that confers unique molecular recognition between polytopic hosts and guests [33,34]. Large hosts with extended cavities are then highly desirable to fully accommodate hydrophobic substrates through multivalency. CD dimers interact more strongly with linear substrates than simple CD receptors owing to their multiple interactions and binding properties. Additionally, the spacer linking the β-CD together supplies additional binding towards the guest molecule upon inclusion complexation. Such cooperative effects were powerful enough to favor the mass transfer in aqueous catalytic systems. RAME-β-CD dimers synthesized by Cu(I)-catalyzed azide–alkyne cycloaddition (CuAAC) proved to be very effective in the Pd-catalyzed Tsuji–Trost reaction [35]. The 'extended' cavity accommodated hydrophobic substrates that were too long to be recognized into a single CD cavity.

Multivalent CD-based polymers were also used as interfacial additives in Rh-catalyzed HF of long-alkyl-chain alkenes (Figure 8) [36]. RAME-β-CD-NH$_2$ was grafted to size-controlled poly(N-acryloyloxysuccinimide) (polyNAS) to give water-soluble CD-based polymers (800 g·L^{-1} at 20 °C) with surface activity (50–60 mN·m^{-1}). Their catalytic performance was assessed in Rh-catalyzed HF of 1-decene and 1-hexadecene. While RAME-β-CD allowed converting 1-decene, polyNAS highly substituted by RAME-β-CD proved to be more efficient than RAME-β-CD to favor the conversion of 1-hexadecene through multivalent interactions. Two close CDs substantially favored the molecular recognition of 1-hexadecene. The local concentration of the latter was significantly increased at the interface, thus favoring contacts with the water soluble rhodium catalyst.

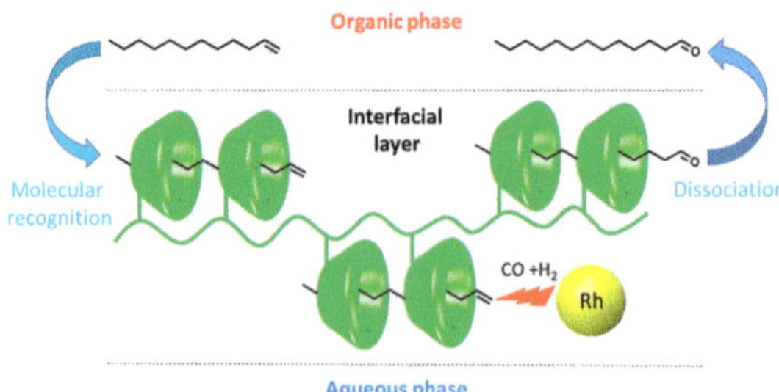

Figure 8. Hydroformylation (HF) of long alkyl chain alkenes using CD-based polymers.

The multivalency concept was then further extended to CD-grafted polyNAS (degree of polymerization of 45) functionalized with the water-soluble sulfonated 2-(diphenylphosphino)ethanamine [37]. The remaining succinimide functions reacted with aminoethanol to give trisubstituted water soluble polymers. Such polymers benefited from both the supramolecular properties of CDs and the capability of phosphanes to coordinate organometallic species. In comparison with sulfonated 2-(diphenylphosphino)ethanamine, the phosphane-grafted, CD-grafted polyNAS gave better activities and aldehyde selectivity in rhodium catalyzed HF of 1-hexadecene at 80 °C under 50 bar of CO/H_2 pressure. As the CD-included substrate and the catalyst were very close, the reaction took place rapidly. For example, 1-hexadecene was fully converted within only 1 h using a 2:1 CD-to-phosphane ratio.

5. Conclusions

Cyclodextrin-based, self-assembly-driven processes have provided compelling evidence that they are powerful tools to improve the catalytic performance of organometallic reactions under biphasic conditions. Such self-assembly-driven processes open fascinating opportunities in the area of organometallic catalysis. The art of self-assembling CDs and ligands, substrates, or other components of the catalytic system promotes both activity and selectivities through molecular recognition and self-organization. The role of CDs in such systems is manifold. They favor contacts between the substrate and the catalyst, modify the aqueous/organic interface, or control the second-sphere of the metal with beneficial effect on the catalytic activity and the selectivities of the reaction. Moreover, CD-based, self-assembly-driven processes appear to be a relevant approach as a rapid screening of the catalytic system can be implemented by just mixing the different components. In this context, we anticipate that CD-based self-assemblies will be used increasingly in the near future to develop more active and more selective catalytic systems. While we still have a long way to fully define the potential of CD-based self-assemblies, the authors hoped this review will inspire the reader to imagine novel approaches involving CD-based, self-assembly-driven processes.

Acknowledgments: The University of Artois is acknowledged for supporting and partially funding this work. Chevreul Institute (FR 2638), Ministère de l'Enseignement Supérieur et de la Recherche, Région Nord—Pas de Calais and FEDER are acknowledged for supporting and funding partially this work.

Author Contributions: F.H. and E.M. wrote the paper equally.

Conflicts of Interest: The authors declare no conflict of interest.

References

1. Raynal, M.; Ballester, P.; Vidal-Ferrana, A.; van Leeuwen, P.W. Supramolecular catalysis. Part 1: Non-covalent interactions as a tool for building and modifying homogeneous catalysts. *Chem. Soc. Rev.* **2014**, *43*, 1660–1733. [CrossRef] [PubMed]
2. Ballester, P.; Vidal-Ferrana, A. *Supramolecular Catalysis*; van Leeuwen, P.W., Ed.; Wiley-VCH: Weinheim, Germany, 2008; pp. 1–27.
3. Catti, L.; Zhang, Q.; Tiefenbacher, K. Advantages of catalysis in self-assembled molecular capsules. *Chem. Eur. J.* **2016**, *22*, 9060–9066. [CrossRef] [PubMed]
4. Leenders, S.H.; Gramage-Doria, R.; de Bruin, B.; Reek, J.N. Transition metal catalysis in confined spaces. *Chem. Soc. Rev.* **2015**, *44*, 433–448. [CrossRef] [PubMed]
5. Bocokić, V.; Kalkan, A.; Lutz, M.; Spek, A.L.; Gryko, D.T.; Reek, J.N. Capsule-controlled selectivity of a rhodium hydroformylation catalyst. *Nat. Commun.* **2013**, *4*, 2670. [CrossRef] [PubMed]
6. Jouffroy, M.; Armspach, D.; Matt, D. Cyclodextrin and phosphorus(III): A versatile combination for coordination chemistry and catalysis. *Dalton Trans.* **2015**, *44*, 12942–12969. [CrossRef] [PubMed]
7. Zaborova, E.; Deschamp, J.; Guieu, S.; Blériot, Y.; Poli, G.; Ménand, M.; Madec, D.; Prestat, G.; Sollogoub, M. Cavitand supported tetraphosphine: Cyclodextrin offers a useful platform for Suzuki-Miyaura cross-coupling. *Chem. Commun.* **2011**, *47*, 9206–9208. [CrossRef] [PubMed]
8. Chevallier, F.; Breit, B. Self-assembled bidentate ligands for Ru-catalyzed anti-Markovnikov hydration of terminal alkynes. *Angew. Chem. Int. Ed.* **2006**, *45*, 1599–1602. [CrossRef] [PubMed]

9. Kuil, M.; Goudriaan, P.E.; Tooke, D.M.; Spek, A.L.; van Leeuwen, P.W.; Reek, J.N. Rigid bis-zinc(II) salphen building blocks for the formation of template-assisted bidentate ligands and their application in catalysis. *Dalton Trans.* **2007**, 2311–2320. [CrossRef] [PubMed]

10. Pignataro, L.; Lynikaite, B.; Cvengroš, J.; Marchini, M.; Piarulli, U.; Gennari, C. Combinations of acidic and basic monodentate binaphtholic phosphites as supramolecular bidentate ligands for enantioselective Rh-catalyzed hydrogenations. *Eur. J. Org. Chem.* **2009**, 2539–2547. [CrossRef]

11. Moteki, S.A.; Takacs, J.M. Exploiting self-assembly for ligand-scaffold optimization: Substrate-tailored ligands for efficient catalytic asymmetric hydroboration. *Angew. Chem. Int. Ed.* **2008**, *47*, 894–897. [CrossRef] [PubMed]

12. Ohmatsu, K.; Ito, M.; Kunieda, T.; Ooi, T. Ion-paired chiral ligands for asymmetric palladium catalysis. *Nat. Chem.* **2012**, *4*, 473–477. [CrossRef] [PubMed]

13. Biedermann, F.; Nau, W.M.; Schneider, H.-J. The hydrophobic effect revisited—Studies with supramolecular complexes imply high-energy water as a noncovalent driving force. *Angew. Chem. Int. Ed.* **2014**, *53*, 11158–11171. [CrossRef] [PubMed]

14. Kataev, E.A.; Müller, C. Recent advances in molecular recognition in water: Artificial receptors and supramolecular catalysis. *Tetrahedron* **2014**, *70*, 137–167. [CrossRef]

15. Machut, C.; Patrigeon, J.; Tilloy, S.; Bricout, H.; Hapiot, F.; Monflier, E. Self-assembled supramolecular bidentate ligands for aqueous organometallic catalysis. *Angew. Chem. Int. Ed.* **2007**, *46*, 3040–3042. [CrossRef] [PubMed]

16. Patrigeon, J.; Hapiot, F.; Canipelle, M.; Menuel, S.; Monflier, E. Cyclodextrin-based supramolecular *P,N* bidentate ligands and their platinum and rhodium complexes. *Organometallics* **2010**, *29*, 6668–6674. [CrossRef]

17. Hapiot, F.; Bricout, H.; Tilloy, S.; Monflier, E. Functionalized cyclodextrins as first and second coordination sphere ligands for aqueous organometallic catalysis. *Eur. J. Inorg. Chem.* **2012**, 1571–1578. [CrossRef]

18. Potier, J.; Guerriero, A.; Menuel, S.; Monflier, E.; Peruzzini, M.; Hapiot, F.; Gonsalvi, L. Cyclodextrins as first and second sphere ligands for Rh(I) complexes of lower-rim PTA derivatives for use as catalysts in aqueous phase hydrogenation. *Catal. Commun.* **2015**, *63*, 74–78. [CrossRef]

19. Menuel, S.; Bertaut, E.; Monflier, E.; Hapiot, F. Cyclodextrin-based PNN supramolecular assemblies: A new class of pincer-type ligands for aqueous organometallic catalysis. *Dalton Trans.* **2015**, *44*, 13504–13512. [CrossRef] [PubMed]

20. Blaszkiewicz, C.; Bricout, H.; Léonard, E.; Len, C.; Landy, D.; Cézard, C.; Djedaïni-Pilard, F.; Monflier, E.; Tilloy, S.A. Cyclodextrin dimer as a supramolecular reaction platform for aqueous organometallic catalysis. *Chem. Commun.* **2013**, *49*, 6989–6991. [CrossRef] [PubMed]

21. Six, N.; Guerriero, A.; Landy, D.; Peruzzini, M.; Gonsalvi, L.; Hapiot, F.; Monflier, E. Supramolecularly controlled surface activity of an amphiphilic ligand. Application to aqueous biphasic hydroformylation of higher olefins. *Catal. Sci. Technol.* **2011**, *1*, 1347–1353. [CrossRef]

22. Leclercq, L.; Schmitzer, A.R. Assembly of tunable supramolecular organometallic catalysts with cyclodextrins. *Organometallics* **2010**, *29*, 3442–3449. [CrossRef]

23. Qi, M.; Tan, P.Z.; Xue, F.; Malhi, H.S.; Zhang, Z.-X.; Young, D.J.; Hor, T.S.A. A supramolecular recyclable catalyst for aqueous Suzuki–Miyaura coupling. *RSC Adv.* **2015**, *5*, 3590–3596. [CrossRef]

24. Yin, Y.; Jiao, S.; Zhang, R.; Wang, X.; Zhang, L.; Yang, L. Construction of a soluble supramolecular glutathione peroxidase mimic based on host-guest interaction. *Curr. Organocatal.* **2015**, *2*, 64–70. [CrossRef]

25. La Sorella, G.; Strukul, G.; Scarso, A. Recent advances in catalysis in micellar media. *Green Chem.* **2015**, *17*, 644–683. [CrossRef]

26. Leclercq, L.; Lacour, M.; Sanon, S.H.; Schmitzer, A.R. Thermoregulated microemulsions by cyclodextrin sequestration: A new approach to efficient catalyst recovery. *Chem. Eur. J.* **2009**, *15*, 6327–6331. [CrossRef] [PubMed]

27. Kairouz, V.; Schmitzer, A.R. Imidazolium-functionalized β-cyclodextrin as a highly recyclable multifunctional ligand in water. *Green Chem.* **2014**, *16*, 3117–3124. [CrossRef]

28. Ferreira, M.; Bricout, H.; Azaroual, N.; Landy, D.; Tilloy, S.; Hapiot, F.; Monflier, E. Cyclodextrin/amphiphilic phosphane mixed systems and their applications in aqueous organometallic catalysis. *Adv. Synth. Catal.* **2012**, *354*, 1337–1346. [CrossRef]

29. Bricout, H.; Léonard, E.; Len, C.; Landy, D.; Hapiot, F.; Monflier, E. Impact of cyclodextrins on the behavior of amphiphilic ligands in aqueous organometallic catalysis. *Beilstein J. Org. Chem.* **2012**, *8*, 1479–1484. [CrossRef] [PubMed]

30. Potier, J.; Menuel, S.; Lyskawa, J.; Fournier, D.; Stoffelbach, F.; Monflier, E.; Woisel, P.; Hapiot, F. Thermoresponsive self-assembled cyclodextrin-end-decorated PNIPAM for aqueous catalysis. *Chem. Commun.* **2015**, *51*, 2328–2330. [CrossRef] [PubMed]

31. Potier, J.; Menuel, S.; Chambrier, M.-H.; Burylo, L.; Blach, J.-F.; Woisel, P.; Monflier, E.; Hapiot, F. Pickering emulsions based on supramolecular hydrogels: Application to higher olefins' hydroformylation. *ACS Catal.* **2013**, *3*, 1618–1621. [CrossRef]

32. Potier, J.; Menuel, S.; Monflier, E.; Hapiot, F. Synergetic effect of randomly methylated β-cyclodextrin and a supramolecular hydrogel in Rh-catalyzed hydroformylation of higher olefins. *ACS Catal.* **2014**, *4*, 2342–2346. [CrossRef]

33. Barnard, A.; Smith, D.K. Self-assembled multivalency: Dynamic ligand arrays for high-affinity binding. *Angew. Chem. Int. Ed.* **2012**, *51*, 6572–6581. [CrossRef] [PubMed]

34. Mulder, A.; Huskens, J.; Reinhoudt, D.N. Multivalency in supramolecular chemistry and nanofabrication. *Org. Biomol. Chem.* **2004**, *7*, 3409–3424. [CrossRef] [PubMed]

35. Six, N.; Menuel, S.; Bricout, H.; Hapiot, F.; Monflier, E. Ditopic cyclodextrin-based receptors: New perspectives in aqueous organometallic catalysis. *Adv. Synth. Catal.* **2010**, *352*, 1467–1475. [CrossRef]

36. Potier, J.; Menuel, S.; Fournier, D.; Fourmentin, S.; Woisel, P.; Hapiot, F.; Monflier, E. Cooperativity in aqueous organometallic catalysis: Contribution of cyclodextrin-substituted polymers. *ACS Catal.* **2012**, *2*, 1417. [CrossRef]

37. Potier, J.; Menuel, S.; Mathiron, D.; Bonnet, V.; Hapiot, F.; Monflier, E. Cyclodextrin-grafted polymers functionalized with phosphanes: A new tool for aqueous organometallic catalysis. *Beilstein J. Org. Chem.* **2014**, *10*, 2642–2648. [CrossRef] [PubMed]

Article

Efficient Degradation of Aqueous Carbamazepine by Bismuth Oxybromide-Activated Peroxide Oxidation

Tuqiao Zhang [1], Shipeng Chu [1], Jian Li [1], Lili Wang [2], Rong Chen [1], Yu Shao [1], Xiaowei Liu [1,3,*] and Miaomiao Ye [1,*]

[1] Institute of Municipal Engineering, College of Civil Engineering and Architecture, Zhejiang University, Hangzhou 310058, China; 0012688@zju.edu.cn (T.Z.); chushipeng@zju.edu.cn (S.C.); xero@zju.edu.cn (J.L.); 21512060@zju.edu.cn (R.C.); shaoyu1979@zju.edu.cn (Y.S.)
[2] Environmental Engineering, Jiyang College, Zhejiang A & F University, Zhuji 311800, China; liilive@163.com
[3] Institute of Port, Coastal and Offshore Engineering, Ocean College, Zhejiang University, Hangzhou 310058, China
* Correspondence: liuxiaowei@zju.edu.cn (X.L.); yemiao008@zju.edu.cn (M.Y.); Tel.: +86-571-8820-8721 (X.L.)

Received: 28 September 2017; Accepted: 18 October 2017; Published: 26 October 2017

Abstract: Bismuth oxyhalide, usually employed as a photocatalyst, has not been tested as an activator of peroxide for water purification. This work explores the potential application of bismuth oxyhalide (BiOX, X = Cl, Br, I)-activated peroxide (H_2O_2; peroxymonosulfate (PMS) and peroxydisulfate) systems for the degradation of carbamazepine (CBZ) in water destined for drinking water. BiOBr showed the highest activity toward the peroxides investigated, especially toward PMS. The most efficient combination, BiOBr/PMS, was selected to further research predominant species responsible for CBZ degradation and toxicity of transformation products. With repeated use of BiOBr, low bismuth-leaching and subtle changes in crystallinity and activity were observed. CBZ degradation was primarily (67.3%) attributable to attack by sulfate radical. Toxicity test and identification of the oxidation products indicated some toxic intermediates may be produced. A possible degradation pathway is proposed. Besides substitution of the hydroxyl groups on the surface of the catalyst particles, PMS's complexation with the lattice Bi(III) through ion exchange with interlayer bromide ion was involved in the decomposition of PMS. The Bi(III)–Bi(V)–Bi(III) redox cycle contributed to the efficient generation of sulfate radicals from the PMS. Our findings provide a simple and efficient process to produce powerful radicals from PMS for refractory pollutant removal.

Keywords: advanced oxidation technologies; bismuth oxybromide; carbamazepine; peroxide; toxicity

1. Introduction

Carbamazepine (CBZ) is a typical pharmaceutically-active compound widely used to control epilepsy and neuropathic pain [1]. Due to its improper disposal after use, CBZ is frequently detected in wastewater treatment plant effluents [2,3], surface water [4] and even drinking water [5] at nanogram- to milligram-per-liter concentrations. CBZ's persistence in the ecosystem has urged researchers to investigate effective treatment technologies. Researchers have previously reported a general removal rate of CBZ below 10% for conventional biological wastewater treatment processes [6]. Even a popular membrane bioreactor removes less than 20% [1,7–9]. Conventional water-treatment processes cannot sufficiently eliminate CBZ [10,11]. More-efficient treatment technologies are urgently needed.

Advanced oxidation technologies (AOTs) have been extensively explored as an alternative solution for degrading and detoxifying bio-recalcitrant and toxic organic pollutants in water. Among AOTs, semiconductor photocatalytic oxidation, which can use UV and visible light energy, has generated much interest [12]. Recently, a relatively new and promising photocatalytic material—bismuth oxyhalide (BiOX, X = Cl, Br, I)—has drawn considerable attention. The bandgap energies for BiOCl,

BiOBr and BiOI are about 3.2, 2.7 and 1.7 eV [13], respectively. Thus, BiOX can be driven by UV/visible light irradiation to degrade pollutants. For liquid-phase photocatalytic BiOX processes, hydroxyl radicals (HO·), superoxide radicals and photoinduced holes have been considered to play important roles [14]. Notably, rare effort has been devoted to producing reactive radicals such as HO· through BiOX-catalyzed decomposition of radical precursors like peroxides.

It is well established that BiOX compounds crystallize in layers in which double slabs of halogen atoms are interleaved with $[Bi_2O_2]$ slabs [13]. Once metal oxides are added into water, dissociative adsorption of water molecules results in the formation of surface hydroxyl groups (–OH) [15]. Peroxides can complex with the metal ions on the surface of metal oxides by replacing part of the surface –OH, generating HO· and $SO_4^{·-}$ radicals through electron transfer [16]. Therefore, besides acting as a photocatalyst, BiOX may activate peroxides and thus induce the degradation of organic pollutants. A preliminary study in our laboratory has found that BiOBr is indeed able to activate peroxymonosulfate (PMS). In addition, peroxides are usually applied as scavengers of photoinduced electrons rather than as radical precursors to increase the degradation rate of organic compounds through a semiconductor photocatalytic processes [17]. To date, CBZ degradation by BiOX-induced peroxide activation has not yet been documented.

In this study, the BiOX-enhanced degradation of CBZ was investigated using peroxide (H_2O_2, PMS and peroxydisulfate (PS)) addition with clean and natural water. Active species contributing to CBZ degradation were identified and those contributions were quantitatively analyzed. A mechanism for active species formation was proposed. Any changes in the biotoxicity of the reaction mixtures during the degradation process were monitored using *Vibrio fisheri* bioassays. Finally, the intermediates involved were characterized using high-performance liquid chromatography coupled with mass spectrometry (HPLC–MS) to elucidate the degradation pathway for CBZ.

2. Results and Discussion

2.1. CBZ Degradation

2.1.1. Degradation of CBZ under Pure Water Background

CBZ adsorbed weakly onto the surface of the three types of BiOX (X = Cl, Br, I) microspheres (Figure 1a), and it was barely oxidized by the peroxides studied (H_2O_2, PS and PMS) alone without the BiOX catalyst (Figure 1b–d). The degradation of CBZ required synergistic action of the peroxides and the BiOX. As shown in Figure 1b,c, the combination of BiOX with H_2O_2 or PS generated only rather slow CBZ degradation. By contrast, the rate with BiOX and PMS was much better (Figure 1d).

Among the three BiOX compounds, BiOBr showed the best degradation effectiveness, exhibiting 100% degradation of the CBZ in less than 5 min. Such catalytic performance was comparable with that of $CuFe_2O_4$ ferrite catalysts (Figure S3, Supplementary Materials). BiOBr was much more efficient than $CuFe_2O_4$ as a PMS activator.

The molecular structure of the peroxides and the physicochemical properties of the BiOX are important to the efficiency of peroxide activation by BiOX. The dissociation energy of the peroxide (O–O) bond in the three peroxides and their ability to bind with BiOX are two key factors. The lower O–O bond energy and easier availability of free persulfate ion under neutral pH compared with H_2O_2 (Table 1) explain why either persulfate (PS or PMS) was easier to activate than H_2O_2 by the same oxyhalide. The observed performance difference between PS and PMS can be attributed to the asymmetrical structure of PMS (compared with the symmetrical structure of H_2O_2 and PS) [18].

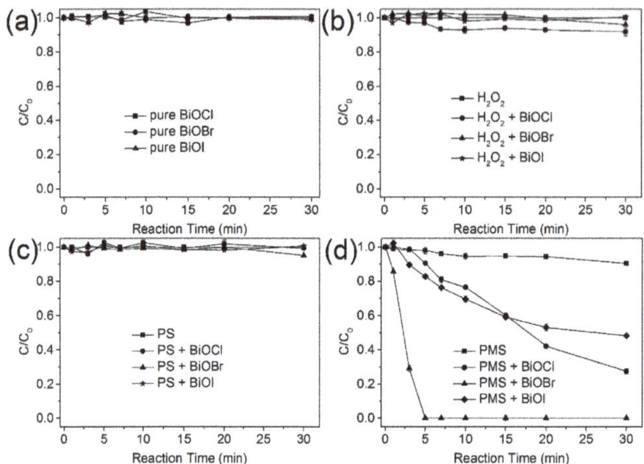

Figure 1. The catalyzed PMS oxidation of CBZ by (**a**) BiOX, (**b**) H_2O_2, (**c**) PS and (**d**) PMS. The initial concentration of CBZ ($[CBZ]_0$) = 5.0 μM, $[H_2O_2]_0 = [PS]_0 = [PMS]_0$ = 4.0 mM, $[BiOX]_0$ = 1.0 g/L, pH = 7.0, T = 24 °C.

Table 1. Physicochemical properties of H_2O_2, PS and PMS.

	Chemical Structure	O–O Bond Energy (E, kJ·mol^{-1})	pKa	Ref.
H_2O_2	H–O–O–H	213.3	$H_2O_2 \rightarrow HO_2^- + H^+$ pKa = 11	[19,20]
PS	^-O_3S–O–O–SO_3^-	140	no change in dissociation form pH > 3	[19]
PMS	H–O–O–SO_3^-	140 < E_{PMS} < 213.3	$HSO_5^- \rightarrow SO_5^{2-} + H^+$ pKa = 9.4	[16,19]

The structure and surface chemistry of the BiOX would of course be expected to affect its activation of peroxides and thus the overall degradation effectiveness. Brunauer Emmett Teller (BET) surface area is one important determinant. The average pore size, particle size and pH_{pzc} were also influential (Table S1, Supplementary Materials). One can find correlation between degradation effectiveness of CBZ and BET surface area of BiOX (Figure S4, Supplementary Materials). The specific surface area of BiOBr was 10% larger than that of BiOCl. That corresponds with more accessible active sites, such as surface hydroxyl groups and Bi(III) ion, which may explain the superior activation with BiOBr. The BiOI had a much-larger specific surface area than the BiOBr, but its activation performance was poorer. That may have reflected the impact of halide released through BiOX hydrolysis (Equation (1), [21]) and ion exchange (Equations (2) and (3), [14]). Interfacial reactions may shed some light on this interesting phenomenon.

$$BiOX\ (X = Cl,\ Br,\ I) + H_2O \xrightarrow{\text{hydrolysis}} BiO^+\text{-}^-OH + X^- + H^+ \tag{1}$$

$$BiOX\ (X = Cl,\ Br,\ I) + HSO_5^- \xrightarrow{\text{ion exchange}} BiO^+\text{-}HSO_5^- + X^- \tag{2}$$

$$2BiOX\ (X = Cl,\ Br,\ I) + S_2O_8^{2-} \xrightarrow{\text{ion exchange}} 2BiO^+\text{-}S_2O_8^{2-} + 2X^- \tag{3}$$

When Cl^-, Br^- or I^- is present in an HO· or $SO_4^{·-}$-based process, halogen radicals such as X· and $X_2^{·-}$ and nonradical X_3^-, HOX and X_2 halogen species with weaker oxidizing power may be formed [22]. Besides its involvement in radical chain reactions with species such as Cl^- and Br^- [22] and direct reaction with peroxides (Equations (4)–(6)), I^- may undergo special reactions with metal ions (Mn^{n+}) when persulfate is present to act as a mediator of electron-transfer reactions (Equations (7)–(10), [23]).

Such reactions will slow the redox cycling of the surface bismuth ions of the BiOI particles (a color change was observed, as shown in Figure S5 of the Supplementary Materials), which would hinder activation of peroxides, tending to cancel out the enhancement caused by a large specific surface area.

$$S_2O_8^{2-} + 2I^- \rightarrow I_2 + 2SO_4^{2-} \quad k = \sim 10^4 \ M^{-1} \cdot s^{-1} \tag{4}$$

$$HSO_5^- + 2I^- + H^+ \rightarrow I_2 + SO_4^{2-} + H_2O \ \text{slow} \tag{5}$$

$$H_2O_2 + 2I^- \rightarrow I_2 + 2OH^- \ \text{slow} \tag{6}$$

$$S_2O_8^{2-} + M^{n+} \rightarrow S_2O_8M^{n-2} \tag{7}$$

$$S_2O_8M^{n-2} + I^- \rightarrow I^+ + M^{n+} + 2SO_4^{2-} \tag{8}$$

$$I^+ + I^- \rightarrow I_2 \tag{9}$$

$$I_2 + H_2O \rightarrow I^- + HIO + H^+ \tag{10}$$

For convenience, the system PMS with BiOBr, which was the most-effective combination of a peroxide and a BiOX, was used in the subsequent stability tests, kinetics studies, investigation of the effects of water quality parameters and elucidation of the CBZ degradation pathway.

2.1.2. Degradation of CBZ under Actual Water Background

To examine CBZ degradation in more realistic conditions, effluent of sand filter from a drinking water plant (ESF), and tap water of Hangzhou (TP) and Qizhen Lake (QZL), were collected. Before the reaction, the water samples were pretreated by static deposition and membrane filtration (0.45 μm filter). The water quality parameters of the samples are listed in Table 2. The pH values were nearly the same, but the other parameters of the various samples were different. Interestingly, the CBZ degradation differed considerably in the different field samples (Figure 2) and from that observed with the MilliQ water. For the tap water, 10 min of reaction was required to achieve complete removal of CBZ. However, the reaction time for complete removal was 20 min when QZL was used, indicating that CBZ degradation was inhibited by some background solutes, presumably inorganic salts and dissolved organic matters (DOM) specific to Qizhen Lake.

Table 2. Water quality of the waters collected from tap water (TP), Qizhen Lake (QZL) and effluent of sand filter (ESF).

	DOM (mg/L)	Bicarbonate (mg/L)	Cl$^-$ (mg/L)	pH	NO$_3^-$-N (mg/L)	UV$_{254}$ (cm^{-1})
TP	3.73	107.06	3.43	7.84	0.24	3.38
QZL	6.61	97.95	2.80	7.83	0.25	4.99
ESF	3.17	100.11	2.36	7.82	0.01	3.10

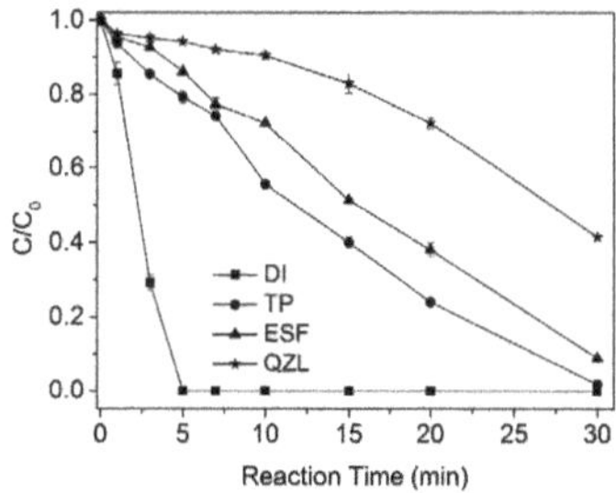

Figure 2. The effects of water matrix on the degradation performance of CBZ. BiOBr dose 0.5 g·L^{-1}, [CBZ]$_0$ = 5 μM, [PMS]$_0$ = 4.0 mM, T = 24 °C.

2.2. Stability of BiOBr as a PMS Activator

To test the durability and stability of the BiOBr catalyst, the particles were reclaimed and reused seven times. The mass-loss during the recycling process was kept within $2.0 \pm 0.5\%$ (Figure 3). The CBZ degradation rate remained almost constant, displaying very stable activation activity for BiOBr with respect to PMS. After seven reaction cycles the specific surface area of the BiOBr had not decreased significantly (virgin, 17.6 $m^2 \cdot g^{-1}$; used, 16.8 $m^2 \cdot g^{-1}$) (data not shown). In addition, the crystallinity of the BiOBr as characterized by XRD showed no apparent change after seven reaction cycles (Figure S6, Supplementary Materials). Moreover, any bismuth ion leaching during the degradation process was found to be minor, not exceeding 120 µg of bismuth ion per gram of BiOBr over 55 min of reaction time (Figure 4). These leaching bismuth ions may result from interaction of non-lattice surface Bi with PMS, given that BiOBr suspension in the absence of PMS almost did not release bismuth ion, and a subtle change of BiOBr crystallinity was observed during repeated use. Generally, this BiOBr catalyst was very stable.

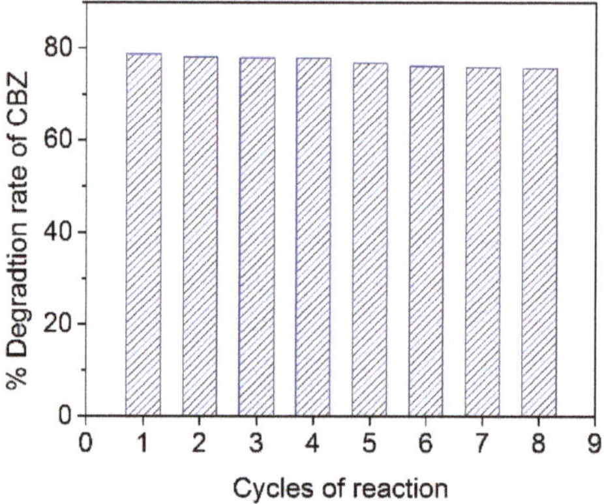

Figure 3. CBZ degradation in repeated experiments with the same BiOBr particles. BiOBr dose 0.5 g·L^{-1}, $[CBZ]_0 = 5$ µM, $[PMS]_0 = 1.0$ mM for all cycles, pH = 7.0, T = 24 °C, reaction for 30 min.

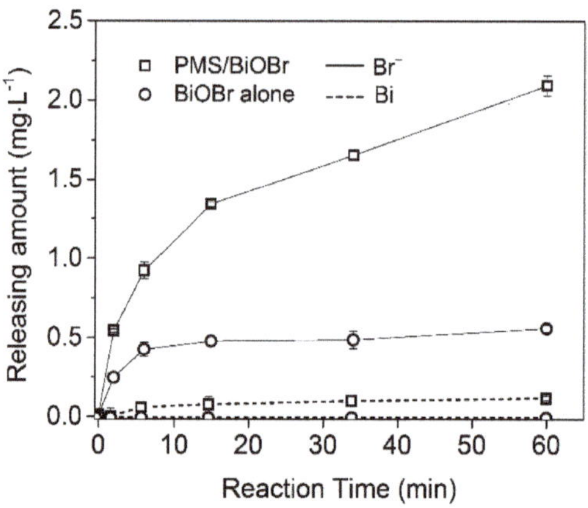

Figure 4. Leaching of Bi and Br^- during the degradation process. BiOBr dose 0.5 g·L^{-1}, $[CBZ]_0 = 5$ µM, $[PMS]_0 = 4.0$ mM, pH = 7.0, T = 24 °C.

2.3. Reactive Species

H_2O_2 and PMS have been reported to generate radicals such as HO· and $SO_4^{·-}$ in the presence of some metal oxides (ferrites [15], iron-cobalt mixed oxide [24] and Co_3O_4 [25]). Since adsorption (Figure 1a), direct oxidation by peroxides (Figure 1b), and leached bismuth ions (Text S4 and Figure S7, Supplementary Materials) make little contribution to CBZ degradation, the CBZ degradation decrease likely resulted from the attack of $SO_4^{·-}$ and/or HO· generated from the activation of PMS by BiOBr. For a direct observation, 5,5-dimethyl-1-pyrroline n-oxide (DMPO) was used to trap the radicals formed in $BiOCl/H_2O_2$ and BiOBr/PMS systems (Figure 5). The hyperfine coupling constants of DMPO radical adducts shown in Figure 5a (a(N) 1.48 mT, a(H) 1.47 mT) are consistent with the assignment of an HO· adduct, proving the formation of HO· in the $BiOCl/H_2O_2$ system. The signal of an $SO_4^{·-}$ adduct (a(N) 1.36 mT, a(H) 1.02 mT, a(H) 0.15 mT, a(H) 0.08 mT, [26]), as well as the signal of an HO· adduct, were detected in the BiOBr/PMS system (Figure 5b), indicating that both $SO_4^{·-}$ and HO· were generated during the reaction. These rapidly degraded CBZ in the BiOBr/PMS system. These results confirm that radical reactions on the BiOBr catalyst–water interface indeed dominated the CBZ degradation in peroxide/BiOX systems.

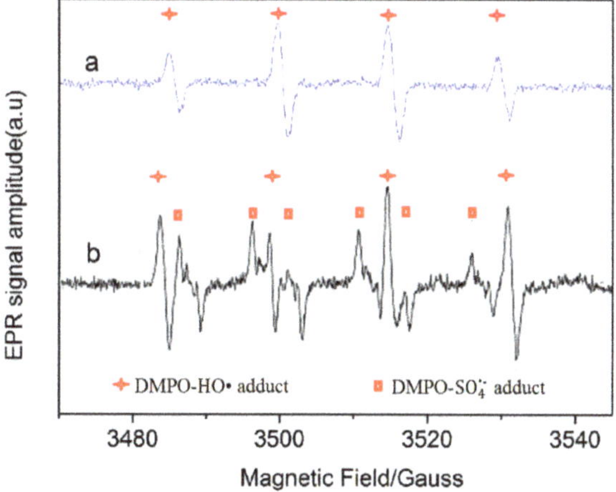

Figure 5. Electron paramagnetic resonance (EPR) spectra obtained from (**a**) $BiOBr/H_2O_2$ and (**b**) BiOBr/PMS systems in the presence of DMPO. BiOBr dose 0.5 g·L^{-1}, [DMPO]$_0$ = 100 mM, [CBZ]$_0$ = 5 μM, [PMS]$_0$ = 4.0 mM, pH = 7.0, T = 24 °C.

The widely-accepted mechanism for $SO_4^{·-}$ formation in PMS heterogeneous catalysis with metal oxides involves the initial substitution of PMS for $-OH$ with the consequent oxidation of metal ions from a low to a higher valence [16]. The high-valence metal ions are then reduced and thus regenerated by PMS. This mechanism requires that the M^{x+}/M^{y+} redox couple of the metal oxide has a higher reduction potential than that of PMS (HSO_5^-).

BiOBr can form surface $-OH$ ($\equiv$ Bi(III)-^-OH_s) and interlayer $-OH$ ($\equiv$ Bi(III)-$^-OH_{inter}$) through dissociative adsorption of water (Equation (11), [15]) and hydrolysis (Equation (12), [21]). This is evidenced by the increasing concentration of bromide in the BiOBr slurry (Figure 3). PMS can undergo direct ion exchange with Br atoms ($-Br$) (Equation (13), [14]) and complexation with Bi atoms in the $[Bi_2O_2]$ layer through substitution of $-OH$ (Equation (14)). It has been reported that surface Bi(III) in BiOCl can be oxidized to a high-valence state [27]. Furthermore, the accumulation of bismuth ion in the solution as the reaction proceeds (Figure 4) and promotion of CBZ degradation by homogeneous Bi(III) (Text S4 and Figure S7, Supplementary Materials) suggest that Bi(III) on the surface of BiOBr particles might be the active sites. Additionally, the reduction potential of Bi(V) to Bi(III) has been reported to be 2.0 V [28], which is higher than that of PMS. Thus, the possible surface reactions generating $SO_4^{\cdot-}$ in the BiOBr/PMS process can be specified as follows:

$$Bi(III)OBr + H_2O \xrightarrow{\text{dissociative adsorption}} Br^- \text{-} [\underset{\underset{OH^- \ H^+}{\uparrow \ \uparrow}}{Bi \ O}]^+ \rightarrow Br^- \text{-} [O(H^+)Bi]^{+} \text{-}^-OH_S \tag{11}$$

$$Br^- \text{-}[O(H^+)Bi]^+ - {}^-OH_s + H_2O \xrightarrow{\text{hydrolysis}} HO^-_{\text{inter}} \text{-}[O(H^+)Bi]^+ \text{-}^-OH_s + Br^- + H^+ \tag{12}$$

$$Br^- \text{-}[O(H^+)Bi]^+ \text{-}^-OH_s + HSO_5^- \rightarrow {}^-SO_3O(HO) \text{-}[O(H^+)Bi]^+ \text{-}OH_s + Br^- \tag{13}$$

$$\equiv Bi(III) \text{-}^-OH_s / {}^-OH_{\text{inter}} + HSO_5^- \rightarrow \equiv Bi(III) \text{-}(HO)OSO_3^- + OH^- \tag{14}$$

$$\equiv Bi(III) \text{-}(HO)OSO_3^- \rightarrow \equiv Bi(IV) \text{-}^-OH_s / {}^-OH_{\text{inter}} + SO_4^{\cdot-} \tag{15}$$

$$\equiv Bi(IV) \text{-}^-OH_s / {}^-OH_{\text{inter}} + HSO_5^- \rightarrow \equiv Bi(IV) \text{-}(HO)OSO_3^- + OH^- \tag{16}$$

$$\equiv Bi(IV) \text{-}(HO)OSO_3^- \rightarrow \equiv Bi(V) \text{-}^-OH_s / {}^-OH_{\text{inter}} + SO_4^{\cdot-} \tag{17}$$

$$\equiv Bi(V) \text{-}^-OH_s / {}^-OH_{\text{inter}} + HSO_5^- \rightarrow \equiv Bi(IV) \text{-}^{\cdot}OOSO_3^- + H_2O \tag{18}$$

$$2 \equiv Bi(IV) \text{-}^{\cdot}OOSO_3^- + 2H_2O \rightarrow 2 \equiv Bi(IV) \text{-}^-OH_s / {}^-OH_{\text{inter}} + 2SO_4^{\cdot-} + O_2 + 2H^+ \tag{19}$$

$$\equiv Bi(IV) \text{-}^-OH_s / {}^-OH_{\text{inter}} + HSO_5^- \rightarrow \equiv Bi(III) \text{-}^{\cdot}OOSO_3^- + H_2O \tag{20}$$

$$2 \equiv Bi(III) \text{-}^{\cdot}OOSO_3^- + 2H_2O \rightarrow 2 \equiv Bi(III) \text{-}^-OH_s / {}^-OH_{\text{inter}} + 2SO_4^{\cdot-} + O_2 + 2H^+. \tag{21}$$

The complexation of BiOBr and PMS was directly confirmed by the ATR–FTIR data (Figure 6). The infrared bands at around 1249 cm^{-1} have been reported as arising from S–O bond stretching vibrations of PMS [29]. This band (peaking at 1257 cm^{-1} in Figure 6a and at 1250 cm^{-1} in Figure 6b) obviously weakened when BiOBr was introduced into the PMS solution. Meanwhile, BiOBr caused the S–O vibration band of the PMS to shift by 7 cm^{-1}; conversely, PMS decreased the band of the surface –OH of BiOBr (peak at 3165 cm^{-1}) (Figure 6c). These two points support Equation (14).

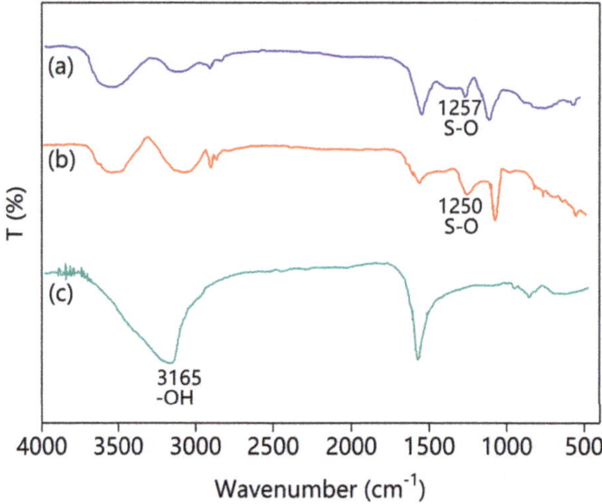

Figure 6. Attenuated total reflectance Fourier transform infrared (ATR–FTIR) spectra of (**a**) mixture of PMS and BiOBr, (**b**) PMS solution and (**c**) BiOBr suspension. Initial pH of the pure water was 5.7; the PMS solution was adjusted to pH 5.7.

The changes in the metal's valence during the reaction were identified by X-ray photoelectron spectroscopy (XPS). As Figure 7a shows, there were two strong peaks (at 159.60 and 164.95 eV) in the Bi region, indicating that Bi(III) was present [30]. Peaks with higher binding energies at 159.69 and 159.77 eV (Figure 7b,c, respectively) can be attributed to Bi(III+x) ions (Bi(IV) or Bi(V)) [27]. The XPS data thus help to elucidate the possible electron-transfer reactions on the surface of the BiOBr particles (Equations (15)–(21)).

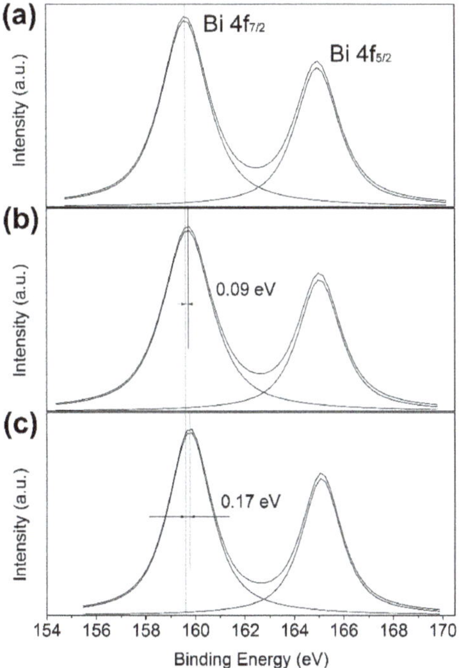

Figure 7. XPS spectra of the Bi4f region for (**a**) BiOBr virgin, (**b**) BiOBr collected after 30-min reaction time and (**c**) BiOBr collected after 60-min reaction time. BiOBr dose 0.5 g·L^{-1}, [CBZ]$_0$ = 5 μM, [PMS]$_0$ = 4.0 mM, pH = 7.0, T = 24 °C.

Two more issues must be addressed to quantitatively explain the excellent performance observed: (1) the reaction rate constants of CBZ with HO· and SO$_4^{·-}$ and (2) the respective contribution of the two radicals.

The CBZ degradation was mainly induced by HO· and SO$_4^{·-}$ attack. These are both strong one-electron oxidants. The kinetics of CBZ degradation are therefore likely to be second-order, as in Equation (22). Equation (23) is the result of the integration of Equation (22) as follows:

$$-\frac{d[\text{CBZ}]}{dt} = \left(k_1 \int [\text{SO}_4^{·-}]dt + \int [\text{HO}·]dt\right)[\text{CBZ}], \tag{22}$$

$$-\ln\frac{[\text{CBZ}]}{[\text{CBZ}]_0} = \left(k_1 \int [\text{SO}_4^{·-}]dt + k_2 \int [\text{HO}·]dt\right)dt, \tag{23}$$

where k$_1$ and k_2 are the reaction rate constants of CBZ with HO· and SO$_4^{·-}$, respectively, and [CBZ]$_0$ and [CBZ] are the CBZ concentrations at the start of the reaction and at time t, respectively. Assuming pseudo-steady state, Equation (23) can be simplified to Equation (24):

$$-\ln\frac{[CBZ]}{[CBZ]_0} = (k_1[SO_4^{\cdot-}]_{ss} + k_2[HO\cdot]_{ss})t = (k_1^* + k_2^*)t, \tag{24}$$

where $[SO_4^{\cdot-}]_{ss}$ and $[HO\cdot]_{ss}$ are the quasi-stationary concentrations of $SO_4^{\cdot-}$ and $HO\cdot$, respectively, and k_1^* and k_2^* are $k_1[SO_4^{\cdot-}]_{ss}$ and $k_2[HO\cdot]_{ss}$, respectively. Thus, the contribution of $SO_4^{\cdot-}$ can be expressed as $k_1^*/(k_1^* + k_2^*)$.

HO· has been reported to attack CBZ with a kinetic constant of 2.05×10^9 $M^{-1}\cdot s^{-1}$ [31], and a second-order rate constant of 1.62×10^9 $M^{-1}\cdot s^{-1}$ can be determined for $SO_4^{\cdot-}$ (Text S5 and Figure S8 in the Supplementary Materials). Both radicals showed nearly diffusion-controlled reactivity with CBZ, which suggests reasonably efficient degradation. The degradation at low initial CBZ concentration ($\leq$5 µM) was well fitted with a pseudo first-order kinetic model, with high values of the correlation coefficient during the initial 7 min of the reaction time (Figure 8). This finding suggests that $k_1^* + k_2^*$ can be assumed to be constant. To determine the contribution of each radical, tertiary butanol (TBA) was added and $k_1^* + k_2^*$ was estimated by fitting a pseudo first-order constant (Figure S9, Supplementary Materials). HO· was considered to be scavenged by the TBA; thus k_2^* can be ignored. $k_1^*/(k_1^* + k_2^*)$ was calculated to be 67.3% ($k_2^* = 0.302$, $k_1^* + k_2^* = 0.449$) at neutral pH. This result tends to confirm that the 67.3% CBZ degradation was due primarily to $SO_4^{\cdot-}$-mediated oxidation.

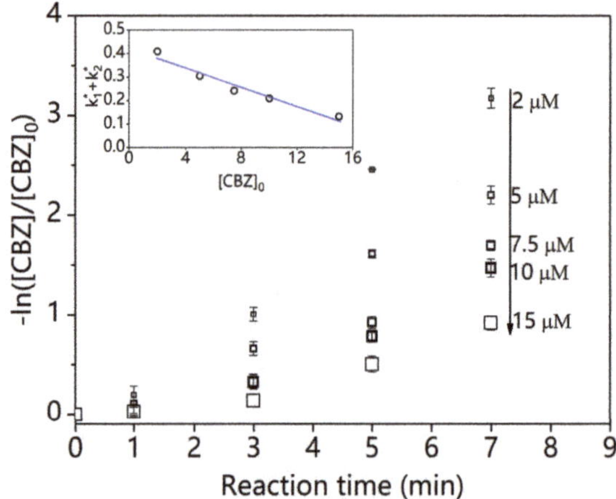

Figure 8. The degradation performance of CBZ by BiOBr/PMS process at different initial concentration of CBZ. BiOBr dose 0.5 g·L^{-1}, [PMS]$_0$ = 4.0 mM, pH = 7.0, T = 24 °C.

2.4. Bromate Formation

Bromate formation in bromide-containing solution in the presence of $SO_4^{\cdot-}$ or HO· has been reported [32]. As for the BiOBr/PMS system, bromide-ion releasing (Figure 4) and radicals ($SO_4^{\cdot-}$ and HO·, Figure 5) have been detected, and bromate formation should be given concern. Therefore, bromate formation during treatment of three actual waters (ESF, TP and QZL) was compared with that of ultrapure water (Figure 9). Within a 30-min reaction time, no detectable bromate formation (<0.002 µg·L^{-1}) for ESF and QZL has been observed, while significant bromate formation happened for ultrapure water (14.3 µM) and TP (1.2 µM). Suppression of bromate formation by natural organic matters in AOTs has been reported [32,33], which is ascribed to scavenging of reactive bromine species and radicals by natural organic matter. A similar effect caused by DOM in ESF, TP and QZL is expected. It is important to highlight that the scavenging of reactive bromine species by natural organic matter or CBZ would pose a risk of yielding organobromated byproducts.

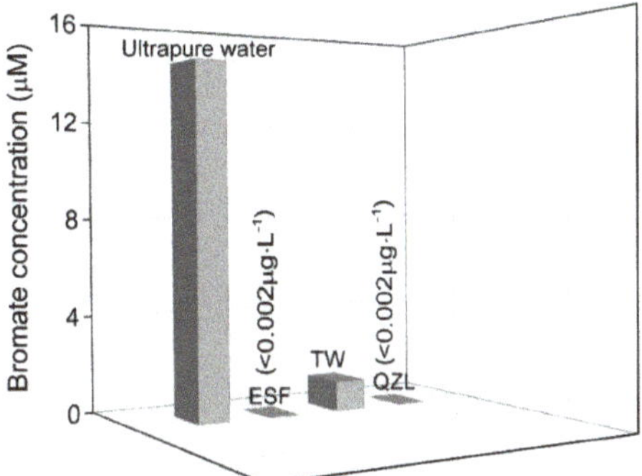

Figure 9. Formation of bromate in three real waters treated by PMS/BiOBr. BiOBr dose 1.0 g·L^{-1}, [CBZ]$_0$ = 5 μM, [PMS]$_0$ = 4.0 mM, pH = 7.0, T = 24 °C.

2.5. Detoxification Performance

The system's acute toxicity was tested using BiOBr as the PMS activator (Figure 10). The toxicity toward *Vibrio fisheri* decreased as the CBZ was eliminated, but over the entire course of the degradation the toxicity decreased more slowly than the CBZ concentration. That suggests that some degradation byproducts also displayed toxicity. Similar to the toxicity, the total organic carbon (TOC) also decreased more slowly than the CBZ concentration. For a 78.2% degradation of CBZ, only 39.7% of the TOC was removed. These results indicate that a long reaction time may be needed to treat CBZ to a safe level. The toxicity evolution can be explained to some extent by exploring CBZ's degradation pathway.

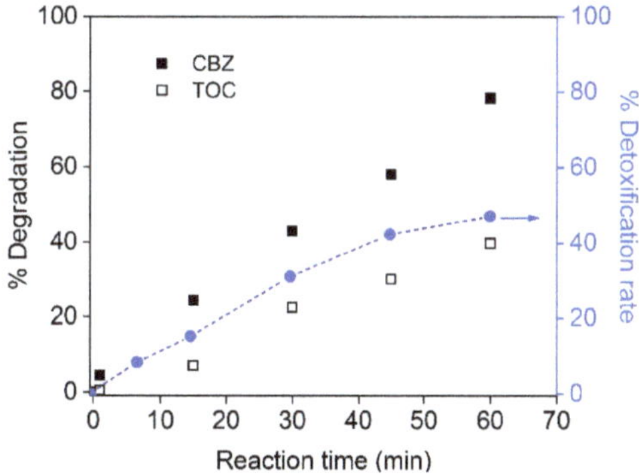

Figure 10. The toxicity evolution of CBZ solution degraded by PMS/BiOBr process. BiOBr dose 0.5 g·L^{-1}, [CBZ]$_0$ = 40 μM, [PMS]$_0$ = 4.0 mM, pH = 7.0, T = 24 °C.

2.6. Degradation Mechanism

To elucidate the mechanism of CBZ degradation, the formation of intermediate products in the degradation of 0.1 mM CBZ in the presence of 4 mM PMS and 0.5 g·L^{-1} BiOBr was detected using high-performance liquid chromatography with tandem mass spectrometric detection (HPLC–MS) (Figure 11). In the figure, the CBZ signal corresponds to the positive ion at m/z 237 generated after H$^+$ addition during the electrospray ionization process. The mass-to-charge peak observed at m/z 259 was contributed by sodium. Meanwhile, intermediates were noted at m/z 180, 196, 210, 224, 239, 251, 253 and 267. Possible molecular structures for these intermediates can be proposed on the basis of the specific *m/z* values, CBZ's chemical structure and the reaction characteristics of $SO_4^{\cdot-}$ and $HO\cdot$. The inferred intermediates suggest the probable degradation pathway for CBZ shown in Figure 12.

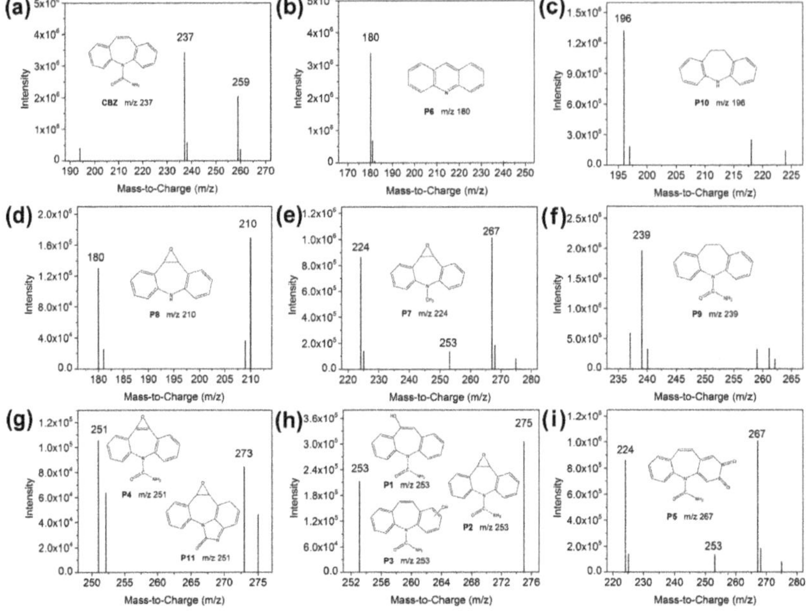

Figure 11. LC–MS spectra of (**a**) CBZ and the possible intermediates; (**b**) *m/z* 180; (**c**) *m/z* 196; (**d**) *m/z* 210; (**e**) *m/z* 224; (**f**) *m/z* 239; (**g**) *m/z* 251; (**h**) *m/z* 253; (**i**) *m/z* 267.

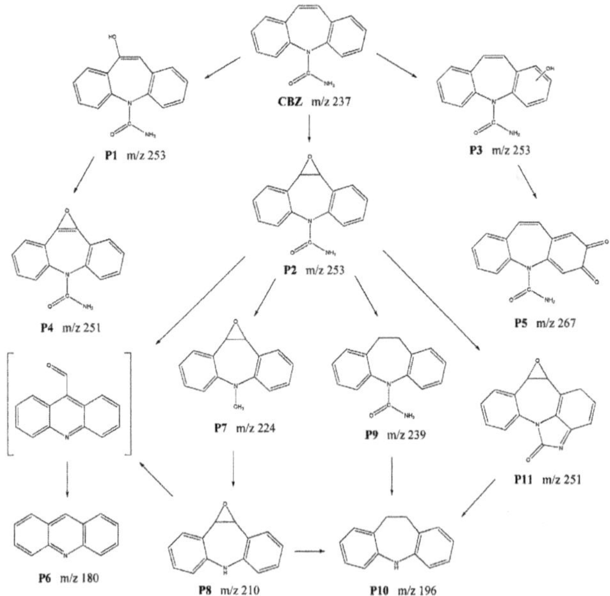

Figure 12. Tentative degradation pathway of CBZ.

The mass peak at 253 was assigned to hydroxy-CBZ (P1 and P3) or 10,11-epoxide-CBZ (P2) on the basis of a report from Sun's group [34]. The further attack of HO· radicals on the hydroxy-CBZ (P1 and P3) induced the formation of 10,11-epoxy-CBZ (P4) and carbamazepine-*o*-quinone (P5) [35,36]. Subsequently, ring contraction in the 10,11-epoxy-CBZ caused the loss of a $-CONH_2$ group. The deamidation produced 9-acridine-9-carboxaldehyde as an intermediate and acridine (P6) as the final form, consistent with previous reports [37,38]. C–N cleavage on the side chains results in the formation of P7 from P2, followed by the formation of P8. Similarly, the ring contraction could also form P11, which could further be degraded to P10. Based on these intermediates, many transformation products retain the functional structure of CBZ and may retain some of its biological toxicity. This effect could partly explain the observed lag in detoxification relative to the degradation of the CBZ parent molecule.

3. Materials and Methods

3.1. Materials and Reagents

Sources of chemicals and reagents are provided in the Supplementary Materials Text S1 . All chemicals were used as received.

3.2. Synthesis and Characterization of BiOX Microstructures

The synthesis of BiOX (X = Cl, Br, I) microstructures was carried out through a solvothermal procedure [14]. This process yielded BiOX microspheres as shown in the scanning electron micrograph which is provided in Figure S1 in the Supplementary Materials. X-ray diffraction (XRD) analysis of a typical BiOBr specimen was conducted using a Rigaku Dmax-2000 diffractometer (Rigaku Co., Tokyo, Japan). A Bruker Optics Matrix-MF spectrometer (Bruker Optics, Ettlingen, Germany) was used for attenuated total-reflection Fourier transform infrared spectroscopy (ATR–FTIR). X-ray photoelectron spectroscopy (XPS) analysis was carried out with an ESCALAB 250Xi spectrometer (Thermo Scientific, Waltham, MA, USA). The specific surface area and average pore size were measured using an ASAP2020HD88 analyzer (Micromeritics, Norcross, GA, USA). The particle-size distribution was quantified using a Mastersizer 2000 laser particle size analyzer (Malvern Instruments Ltd., Malvern, UK). Acid–base titration was employed to determine the point of zero charge (pH_{pzc}).

3.3. Experimental Procedures

The CBZ degradation experiments were performed in a room thermostated 24 °C and under dark conditions. The 0.02 mM CBZ stock solution was stirred for 30 min before use. Except for the best peroxide-activator screening experiments where 1 mM peroxide (PS, PMS or H_2O_2) was used, the CBZ concentration was set at 5 μM and the initial concentration of peroxide (PS, PMS or H_2O_2) was kept at 4 mM in the reaction solution. Subsequently, 0.10 g of the BiOX microspheres was dispersed into 200 mL of the reaction solution, and the pH was adjusted with 10 mM Michaelis buffer (to pH 4.0 or 5.0) or borate buffer (to pH 7.0, 9.0 or 11.0) as necessary. The anions (tartrate ion and tetraborate ion) in the buffer showed weak ion exchange (Figure S2, Supplementary Materials), which means they had little influence on the BiOX-catalyzed peroxide oxidation. During the reaction, samples were collected using a syringe at predetermined time intervals and transferred to vials containing a quencher (100 μL of 0.2 M $NaNO_2$ [16]). The mixture was then filtered immediately for analysis of the CBZ concentration and to evaluate any toxicity.

3.4. Analysis

The CBZ content in the samples was determined using an Agilent 1200 HPLC (Agilent, Palo Alto, CA, USA) equipped with a UV–vis detector. The degradation intermediates were concentrated on a Gilson GX-271 ASPEC apparatus (Gilson, Middleton, WI, USA) and analyzed with an Agilent 6460 triple-quad HPLC–MS (Agilent, Palo Alto, CA, USA). The release of bromide (Br^-) from the BiOBr materials and possible bromate formation were monitored using a Dionex ICS-2000 ion chromatograph

(Dionex, Sunnyvale, CA, USA). The bismuth leaching during the degradation process was detected using a PE NexION 300Q inductively-coupled plasma mass spectrometer (Perkin Elmer, Oak Brook, IL, USA). The detailed parameters are shown in Text S2 of the Supplementary Materials.

4. Conclusions

This study evaluated catalytic peroxide (PMS, PS and H_2O_2) oxidation of CBZ using BiOX as the catalyst. The results show that the BiOX microsphere catalyst activated PMS most efficiently. BiOBr generated the best kinetics and it was found to be both efficient and stable for CBZ degradation. Sulfate radical was determined to be the primary oxidative species in the BiOBr/PMS system. The high reduction potential of Bi(V)/Bi(III) facilitates the efficient formation of the sulfate radical. The degradation intermediates of CBZ retain some biotoxicity and a relatively long reaction time may be necessary to detoxify CBZ completely. Bromide is released from BiOBr in the process through hydrolysis and ion exchange, so bromate produced by the radical oxidation of bromide must be examined. Presence of dissolved organic matter acts as an inhibitor of bromate formation. Deposit of BiOBr onto materials with strong ion-exchange ability to locate bromide ions may relieve such concern. In addition, bismuth leaching, although very slow, is expected to shorten the catalyst's lifespan. Modifications, such as metal doping, may be necessary to improve the stability and recyclability of BiOBr.

Supplementary Materials: The following are available online at www.mdpi.com/2073-4344/7/11/315/s1, Text S1: Chemicals, Text S2: Parameters of analysis, Text S3: Toxicity evaluation of samples, Text S4: Effect of leaching bismuth ion, Text S5: Determination of $k(SO_4^- + CBZ)$, Figure S1: SEM images of BiOX microspheres: (a) BiOCl, (b) BiOBr, (c) BiOI, Figure S2: Effect of anions in buffer on the releasing of bromide ion, Figure S3: Comparison of CBZ degradation by PMS/BiOBr and PMS/CuFe$_2$O$_4$, Figure S4: Normalized degradation rate of CBZ based on specific surface area for PMS/BiOBr, Figure S5: Color change of BiOI suspension during the degradation process, Figure S6: XRD spectra of BiOBr particles before and after the repeated catalytic PMS oxidation (7 cycles), Figure S7. Effect of leaching bismuth ion on the degradation of CBZ, Figure S8: -ln ([CBZ]/[CBZ]$_0$) vs. -ln ([pCBA]/[pCBA]$_0$), Figure S9: CBZ degradation in the presence or absence of TBA, Table S1: BET surface area, average pore and particle size, and pHpzc of BiOX (X = Cl, Br, I).

Acknowledgments: This work was financially supported by the National Natural Science Foundation of China (Grant No. 51408539 and 51478417), the Public Projects of Zhejiang Province (Grant No. 2017C33174), the Major International (Regional) Joint Research Program of China (No.51761145022), and the National Key Research and Development Program of China (No. 2016YFC0400601, 2016YFC0400606). The authors also greatly appreciate the additional support from Jiyang College of Zhejiang A & F University (Grant No. 04251700010) and the special S&T project on the treatment and control of water pollution (Grant No. 2017ZX07201-003).

Author Contributions: X.L. and M.Y. conceived and designed the experiments; T.Z., S.C., J.L., and L.W. performed the experiments and analyzed the data; R.C. and Y.S. contributed reagents/materials/analysis tools; L.W. wrote the paper.

Conflicts of Interest: The authors declare no conflict of interest.

References

1. Miao, X.; Metcalfe, C.D. Determination of carbamazepine and its metabolites in aqueous samples using liquid chromatography−electrospray tandem mass spectrometry. *Anal. Chem.* **2003**, *75*, 3731–3738. [CrossRef] [PubMed]

2. Ternes, T.A. Occurrence of drugs in german sewage treatment plants and rivers. *Water Res.* **1998**, *32*, 3245–3260. [CrossRef]

3. Heberer, T. Tracking persistent pharmaceutical residues from municipal sewage to drinking water. *J. Hydrol.* **2002**, *266*, 175–189. [CrossRef]

4. Ginebreda, A.; Muñoz, I.; de Alda, M.L.; Brix, R.; López-Doval, J.; Barceló, D. Environmental risk assessment of pharmaceuticals in rivers: Relationships between hazard indexes and aquatic macroinvertebrate diversity indexes in the Llobregat River (NE Spain). *Environ. Int.* **2010**, *36*, 153–162. [CrossRef] [PubMed]

5. Benotti, M.J.; Trenholm, R.A.; Vanderford, B.J.; Holady, J.C.; Stanford, B.D.; Snyder, S.A. Pharmaceuticals and endocrine disrupting compounds in us drinking water. *Environ. Sci. Technol.* **2008**, *43*, 597–603. [CrossRef]

6. Zhang, Y.; Geißen, S.U.; Gal, C. Carbamazepine and diclofenac: Removal in wastewater treatment plants and occurrence in water bodies. *Chemosphere* **2008**, *73*, 1151–1161. [CrossRef] [PubMed]
7. Clara, M.; Strenn, B.; Kreuzinger, N. Carbamazepine as a possible anthropogenic marker in the aquatic environment: Investigations on the behaviour of carbamazepine in wastewater treatment and during groundwater infiltration. *Water Res.* **2004**, *38*, 947–954. [CrossRef] [PubMed]
8. Joss, A.; Keller, E.; Alder, A.C.; Göbel, A.; McArdell, C.S.; Ternes, T.; Siegrist, H. Removal of pharmaceuticals and fragrances in biological wastewater treatment. *Water Res.* **2005**, *39*, 3139–3152. [CrossRef] [PubMed]
9. Radjenovic, J.; Petrovic, M.; Barceló, D. Analysis of pharmaceuticals in wastewater and removal using a membrane bioreactor. *Anal. Bioanal. Chem.* **2007**, *387*, 1365–1377. [CrossRef] [PubMed]
10. Carballa, M.; Omil, F.; Lema, J.M.; Llompart, M.; Garcia-Jares, C.; Rodriguez, I.; Gómez, M.; Ternes, T. Behavior of pharmaceuticals, cosmetics and hormones in a sewage treatment plant. *Water Res.* **2004**, *38*, 2918–2926. [CrossRef] [PubMed]
11. Suárez, S.; Carballa, M.; Omil, F.; Lema, J.M. How are pharmaceutical and personal care products (PPCPs) removed from urban wastewaters? *Rev. Environ. Sci. Biotechnol.* **2008**, *7*, 125–138. [CrossRef]
12. Wang, X.; Yang, W.; Li, F.; Zhao, J.; Liu, R.; Liu, S.; Li, B. Construction of amorphous TiO_2/BiOBr heterojunctions via facets coupling for enhanced photocatalytic activity. *J. Hazard. Mater.* **2015**, *292*, 126–136. [CrossRef] [PubMed]
13. Cheng, H.; Huang, B.; Dai, Y. Engineering BiOX (X = Cl, Br, I) nanostructures for highly efficient photocatalytic applications. *Nanoscale* **2014**, *6*, 2009–2026. [CrossRef] [PubMed]
14. Li, J.; Sun, S.; Qian, C.; He, L.; Chen, K.; Zhang, T.; Chen, Z.; Ye, M. The role of adsorption in photocatalytic degradation of ibuprofen under visible light irradiation by BiOBr microspheres. *Chem. Eng. J.* **2016**, *297*, 139–147. [CrossRef]
15. Guan, Y.; Ma, J.; Ren, Y.; Liu, Y.; Xiao, J.; Lin, L.; Zhang, C. Efficient degradation of atrazine by magnetic porous copper ferrite catalyzed peroxymonosulfate oxidation via the formation of hydroxyl and sulfate radicals. *Water Res.* **2013**, *47*, 5431–5438. [CrossRef] [PubMed]
16. Zhang, T.; Zhu, H.; Croué, J.P. Production of sulfate radical from peroxymonosulfate induced by a magnetically separable $CuFe_2O_4$ spinel in water: Efficiency, stability, and mechanism. *Environ. Sci. Technol.* **2013**, *47*, 2784–2791. [CrossRef] [PubMed]
17. Wang, Y.; Hong, C. Effect of hydrogen peroxide, periodate and persulfate on photocatalysis of 2-chlorobiphenyl in aqueous TiO_2 suspensions. *Water Res.* **1999**, *33*, 2031–2036. [CrossRef]
18. Rastogi, A.; Al-Abed, S.R.; Dionysiou, D.D. Effect of inorganic, synthetic and naturally occurring chelating agents on Fe(II) mediated advanced oxidation of chlorophenols. *Water Res.* **2009**, *43*, 684–694. [CrossRef] [PubMed]
19. Yang, S.; Wang, P.; Yang, X.; Shan, L.; Zhang, W.; Shao, X.; Niu, R. Degradation efficiencies of azo dye acid orange 7 by the interaction of heat, uv and anions with common oxidants: Persulfate, peroxymonosulfate and hydrogen peroxide. *J. Hazard. Mater.* **2010**, *179*, 552–558. [CrossRef] [PubMed]
20. Von Sonntag, C. Advanced oxidation processes: Mechanistic aspects. *Water Sci. Technol.* **2008**, *58*, 1015–1021. [CrossRef] [PubMed]
21. Taylor, P.; Lopata, V.J. Some phase relationships between basic bismuth chlorides in aqueous solutions at 25 °C. *Can. J. Chem.* **1987**, *65*, 2824–2829. [CrossRef]
22. Yang, Y.; Pignatello, J.J.; Ma, J.; Mitch, W.A. Comparison of halide impacts on the efficiency of contaminant degradation by sulfate and hydroxyl radical-based advanced oxidation processes (AOPs). *Environ. Sci. Technol.* **2014**, *48*, 2344–2351. [CrossRef] [PubMed]
23. House, D.A. Kinetics and mechanism of oxidations by peroxydisulfate. *Chem. Rev.* **1962**, *62*, 185–203. [CrossRef]
24. Yang, Q.; Choi, H.; Al-Abed, S.R.; Dionysiou, D.D. Iron–cobalt mixed oxide nanocatalysts: Heterogeneous peroxymonosulfate activation, cobalt leaching, and ferromagnetic properties for environmental applications. *Appl. Catal. B Environ.* **2009**, *88*, 462–469. [CrossRef]
25. Anipsitakis, G.P.; Stathatos, E.; Dionysiou, D.D. Heterogeneous activation of oxone using Co_3O_4. *J. Phys. Chem. B* **2005**, *109*, 13052–13055. [CrossRef] [PubMed]
26. Furman, O.S.; Teel, A.L.; Watts, R.J. Mechanism of base activation of persulfate. *Environ. Sci. Technol.* **2010**, *44*, 6423–6428. [CrossRef] [PubMed]

27. Zhang, L.; Wang, W.; Jiang, D.; Gao, E.; Sun, S. Photoreduction of CO_2 on BiOCl nanoplates with the assistance of photoinduced oxygen vacancies. *Nano Res.* **2015**, *8*, 821–831. [CrossRef]

28. Bard, A.J.; Parsons, R.; Jordan, J. *Standard Potentials in Aqueous Solution*; Marcel Dekker, Inc.: New York, NY, USA, 1985; pp. 180–187.

29. Gonzalez, J.; Torrent-Sucarrat, M.; Anglada, J.M. The reactions of SO_3 with HO_2 radical and H_2O ... HO_2 radical complex. Theoretical study on the atmospheric formation of HSO_5 and H_2SO_4. *Phys. Chem. Chem. Phys.* **2010**, *12*, 2116–2125. [CrossRef] [PubMed]

30. Ye, L.; Jin, X.; Liu, C.; Ding, C.; Xie, H.; Chu, K.; Wong, P. Thickness-ultrathin and bismuth-rich strategies for BiOBr to enhance photoreduction of CO_2 into solar fuels. *Appl. Catal. B Environ.* **2016**, *187*, 281–290. [CrossRef]

31. Vogna, D.; Marotta, R.; Andreozzi, R.; Napolitano, A.; d'Ischia, M. Kinetic and chemical assessment of the UV/H_2O_2 treatment of antiepileptic drug carbamazepine. *Chemosphere* **2004**, *54*, 497–505. [CrossRef]

32. Lutze, H.V.; Bakkour, R.; Kerlin, N.; von Sonntag, C.; Schmidt, T.C. Formation of bromate in sulfate radical based oxidation: Mechanistic aspects and suppression by dissolved organic matter. *Water Res.* **2014**, *53*, 370–377. [CrossRef] [PubMed]

33. Fang, J.; Shang, C. Bromate formation from bromide oxidation by the UV/persulfate process. *Environ. Sci. Technol.* **2012**, *46*, 8976–8983. [CrossRef] [PubMed]

34. Guan, Y.; Ma, J.; Li, X.; Fang, J.; Chen, L. Influence of pH on the formation of sulfate and hydroxyl radicals in the UV/peroxymonosulfate system. *Environ. Sci. Technol.* **2011**, *45*, 9308–9314. [CrossRef] [PubMed]

35. Monsalvo, V.M.; Lopez, J.; Munoz, M.; de Pedro, Z.M.; Casas, J.A.; Mohedano, A.F.; Rodriguez, J.J. Application of Fenton-like oxidation as pre-treatment for carbamazepine biodegradation. *Chem. Eng. J.* **2015**, *264*, 856–862. [CrossRef]

36. Rao, Y.F.; Qu, L.; Yang, H.; Chu, W. Degradation of carbamazepine by Fe(II)-activated persulfate process. *J. Hazard. Mater.* **2014**, *268*, 23–32. [CrossRef] [PubMed]

37. Daghrir, R.; Drogui, P.; Dimboukou-Mpira, A.; El Khakani, M. Photoelectrocatalytic degradation of carbamazepine using Ti/TiO_2 nanostructured electrodes deposited by means of a pulsed laser deposition process. *Chemosphere* **2013**, *93*, 2756–2766. [CrossRef] [PubMed]

38. Tran, N.; Drogui, P.; Zaviska, F.; Brar, S.K. Sonochemical degradation of the persistent pharmaceutical carbamazepine. *J. Environ. Manag.* **2013**, *131*, 25–32. [CrossRef] [PubMed]

Review

Metal-Catalyzed Intra- and Intermolecular Addition of Carboxylic Acids to Alkynes in Aqueous Media: A Review

Javier Francos and Victorio Cadierno *

Laboratorio de Compuestos Organometálicos y Catálisis (Unidad Asociada al CSIC), Centro de Innovación en Química Avanzada (ORFEO-CINQA), Departamento de Química Orgánica e Inorgánica, IUQOEM, Facultad de Química, Universidad de Oviedo, Julián Clavería 8, E-33006 Oviedo, Spain; francosjavier@uniovi.es
* Correspondence: vcm@uniovi.es; Tel.: +34-985-103453

Received: 19 October 2017; Accepted: 2 November 2017; Published: 6 November 2017

Abstract: The metal-catalyzed addition of carboxylic acids to alkynes is a very effective tool for the synthesis of carboxylate-functionalized olefinic compounds in an atom-economical manner. Thus, a large variety of synthetically useful lactones and enol-esters can be accessed through the intra- or intermolecular versions of this process. In order to reduce the environmental impact of these reactions, considerable efforts have been devoted in recent years to the development of catalytic systems able to operate in aqueous media, which represent a real challenge taking into account the tendency of alkynes to undergo hydration in the presence of transition metals. Despite this, different Pd, Pt, Au, Cu and Ru catalysts capable of promoting the intra- and intermolecular addition of carboxylic acids to alkynes in a selective manner in aqueous environments have appeared in the literature. In this review article, an overview of this chemistry is provided. The synthesis of β-oxo esters by catalytic addition of carboxylic acids to terminal propargylic alcohols in water is also discussed.

Keywords: aqueous catalysis; alkynes; carboxylic acids; alkynoic acids; hydrocarboxylation reactions; cycloisomerization reactions; enol esters; lactones; β-oxo esters

1. Introduction

The transition metal-catalyzed heterofunctionalization of alkynes by addition of nucleophiles to the C≡C bond has emerged in recent years as a versatile synthetic tool in organic chemistry [1–3]. In particular, the addition of carboxylic acids to alkynes, reaction also referred in the literature as hydro-oxycarbonylation or hydrocarboxylation of alkynes, represents a straightforward way to obtain different types of olefinic esters with atom economy [4–6]. Thus, the intramolecular version of the process, i.e., the cycloisomerization of alkynoic acids, produces unsaturated lactones (Scheme 1) which are common structural motifs found in natural products and biologically active molecules, as well as valuable synthetic intermediates [7–10]. A large variety of transition metal complexes have been used to catalyze these reactions through the π-activation of the carbon-carbon triple bond of the alkynoic acid, the regioselectivity (*exo* or *endo* addition of the carboxylate to the C≡C bond leading to lactones **A** or **B**, respectively) and stereoselectivity of the process being dependent on the metal employed, the length of the hydrocarbon chain connecting the acid and alkyne units, and the terminal or internal nature of the alkyne functionality [1–6]. Among the most commonly employed metals are Pd, Ag, Au and Rh, including examples of heterogeneous systems [11], which have shown a high efficacy for the selective preparation of 5-, 6-, and to a lesser extent, 7-membered ring lactones [1–6].

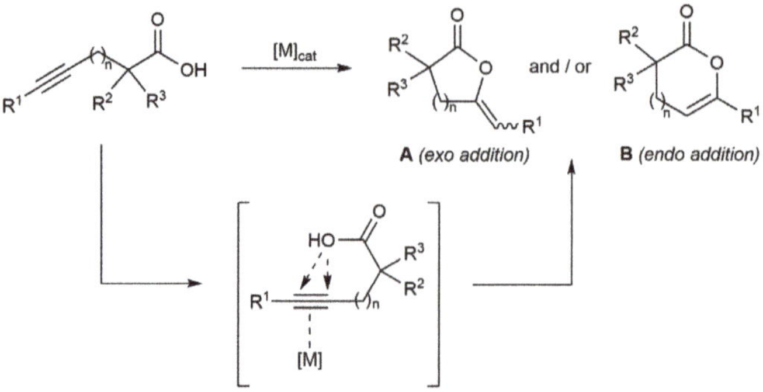

Scheme 1. The catalytic cycloisomerization of alkynoic acids.

The intermolecular addition of carboxylic acids to alkynes also presents enormous interest since the reaction products, i.e., enol esters (**C** or **D** in Scheme 2), are versatile building blocks in organic synthesis and material science. For example, to name just a few of their myriad applications, they are widely employed as mild acylating reagents [12,13], as monomers in diverse polymerization reactions [14–16], as substrates in asymmetric hydrogenation for the generation of enantioenriched alcohols [17–19], or as starting materials for different cross-coupling processes [20–22]. As for the cycloisomerization of alkynoic acids, a huge number of catalytic systems involving late transition metals have been described, with those based on ruthenium playing a prominent role [1–5]. However, for a long time, this reaction was limited to terminal alkynes, since most catalysts failed to enable the hydrocarboxylation of internal alkynes as a consequence of their greater steric hindrance, which disfavours their coordination to the metal catalyst. At this point, it should be noted that a reduced reactivity of internal vs. terminal C≡C bonds is also commonly observed in the cyclization reactions of alkynoic acids, although in this case the problem is not so marked as the intramolecular process is much more favoured from a thermodynamic point of view. Fortunately, recent works focused in the design of increasingly active catalysts have made possible to overcome this limitation and some Ru-, Pd-, Ag- and Au-based systems active with internal alkynes are now available [23].

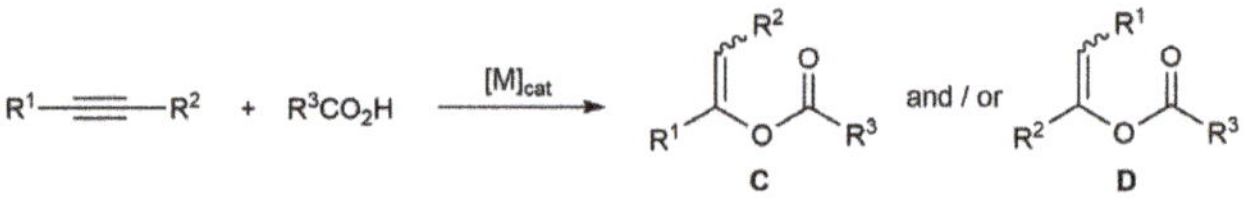

Scheme 2. The intermolecular addition of carboxylic acids to alkynes.

On the other hand, the increasing awareness of environmental concerns has stimulated the development of metal-catalyzed reactions in aqueous media, since water is cheaper, safer and benigner compared to the traditional petroleum-derived organic solvents [24–26]. In addition, the use of water (or aqueous biphasic mixtures) allows in many cases an easy catalyst/product separation, thus allowing the effective recycling of the catalytically active species, which is another key aspect in the Green Chemistry context [27]. However, despite the growing interest in aqueous catalysis, the use of this environmentally benign solvent in the hydrocarboxylation of alkynes been for long time neglected, probably due to the concerns of a competing hydration of the alkyne substrates to form carbonyl compounds, a process that is also catalyzed by transition metals [1–6,28]. In fact, in a general review on the "catalytic reactions of alkynes in water", published by Chen and Li in 2006, no examples of hydrocarboxylation reactions were collected [28]. Catalytic systems able to operate selectively in aqueous environments have only appeared in the literature in recent years, and are the subject of the present review article. Intra- and intermolecular processes are covered, including examples of sequential transformations of synthetic interest. The access to β-oxo esters by the catalytic addition of carboxylic acids to terminal propargylic alcohols in water is also discussed.

2. Intramolecular Processes

2.1. Cycloisomerization of Preformed or In Situ Generated Alkynoic Acids

The first example of such reactions in an aqueous environment was described by Hidai and co-workers in 1996 employing the mixed-metal cubane-type cluster complex [PdMo$_3$S$_4$(tacn)$_3$Cl][PF$_6$]$_3$ (**1**; tacn = 1,4,7-triazacyclononane) as catalyst [29]. Thus, they found that, in combination with NEt$_3$, **1** was able to transform dipropargylmalonic acid **2** into the enol lactone **3** in water at room temperature (Scheme 3). Unfortunately, although the rate and yield of the reaction were very similar to those observed in acetonitrile, this latter solvent was selected by the authors to study the scope of the process. In this regard, different 5-, 6- and 7-membered ring lactones could be obtained from the corresponding alkynoic acids with complex **1**, which showed an enhanced reactivity in comparison with classical mononuclear Pd(II) sources, such as [PdCl$_2$(PhCN)$_2$] or Na$_2$[PdCl$_4$]. In line with this, despite the presence of three molybdenum atoms in the structure of **1**, the catalytic reactions were assumed to proceed only at the palladium center.

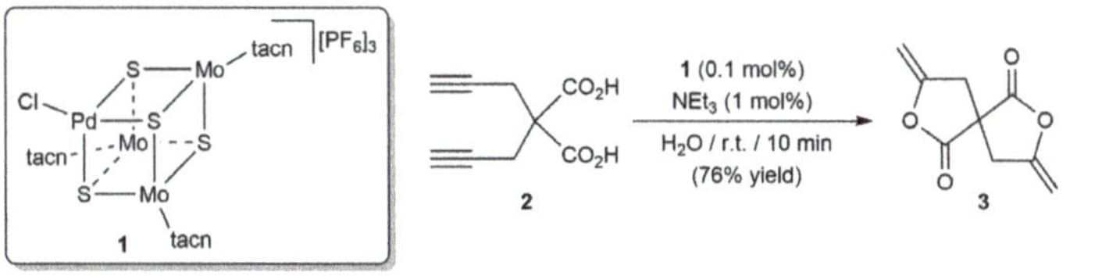

Scheme 3. Cyclosiomerization of dipropargylmalonic acid **2** in water using a cluster catalyst.

More recently, other palladium-based catalysts for the cyclization of alkynoic acids in aqueous media have been described. In particular, García-Álvarez and co-workers disclosed that transformation of γ-alkynoic acids into five-membered ring enol-lactones (5-*exo-dig* cyclization) can be conveniently and selectively achieved in pure water, and under aerobic conditions, by using catalytic amounts of the dinuclear Pd(II) derivative *trans*-[PdCl$_2$\{μ^2-N,S-(PTA)=NP(=S)(OEt)$_2$\}]$_2$ (**4**) (Scheme 4) [30]. This catalyst features a bridging iminophosphorane-type ligand derived from the well-known hydrophilic phosphine PTA (1,3,5-triaza-7-phosphaadamantane) [31,32], which facilitates its solubility in water.

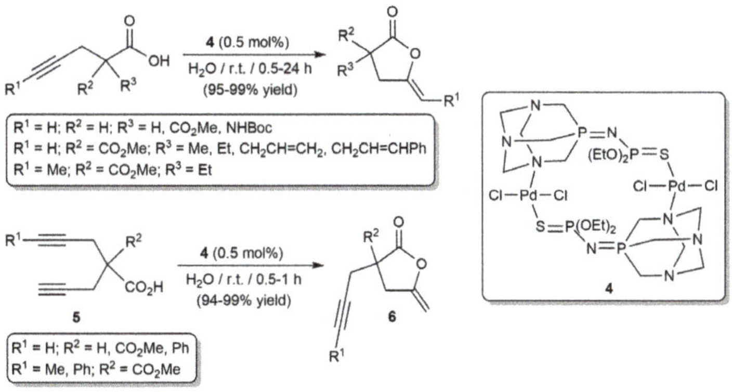

Scheme 4. Cycloisomerization of different γ-alkynoic acids in water catalyzed by a dinuclear Pd(II) complex.

As shown in Scheme 4, the reactions proceeded cleanly at r.t. (room temperature) without formation of any by-product derived from the hydration of the C≡C bond of the substrates or from the hydrolysis of the (Z)-γ-alkylidene butyrolactone products. The process was operative with γ-alkynoic acids containing both terminal and internal C≡C bonds, although for the latter a much longer reaction time was needed (24 instead of 0.5–2 h). It is also worth noting that (*i*) this catalytic system could be recycled up to 10 consecutive runs for the cyclization of the model 4-pentynoic acid (cumulative TON (turnover number) of 982), and (*ii*) it could be applied in the desymmetrization of bispropargylic carboxylic acids **5**, affording enol-lactones **6** containing an intact propargylic side arm in excellent yields.

Although in a limited number of examples (only six substrates were explored), the group of Nakamura also demonstrated the capacity of the NCN-pincer Pd(II) derivative **7** (Figure 1) to promote the high-yield formation of five-membered ring lactones from both γ- and β-alkynoic acids (5-*exo-dig* and 5-*endo-dig* cyclization, respetively) in pure water [33]. Complex **7** easily generates hydrogen-bonding-based supramolecular gels, with well-organized Pd-arrays, in organic solvents. For the catalytic reactions, which were performed at 70 °C and in the presence of Et₃N (3 mol %), a water-insoluble xerogel prepared from **7** in toluene was employed (0.5 mol % Pd content). Interestingly, the Pd-gel catalyst could be recovered by filtration and reused three times in the cyclization of 4-pentynoic acid without loss of catalytic activity. However, it should also be mentioned that, contrary to the case of the dinuclear Pd(II) complex **4**, when a bispropargylic carboxylic acid was employed as substrate, partial hydration of propargylic side arm of the enol-lactone product was observed.

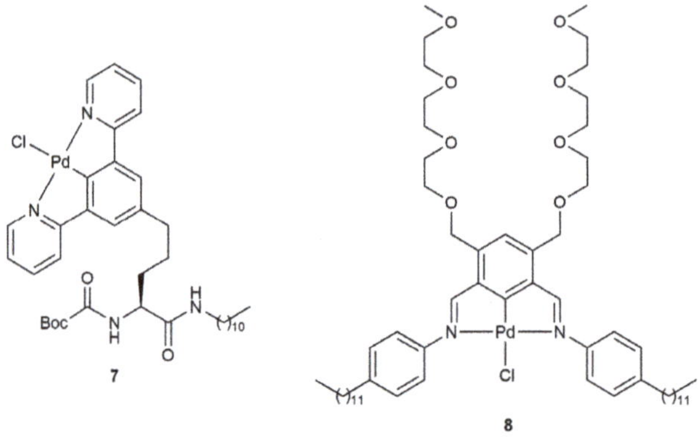

Figure 1. Structure of the pincer palladium(II) complexes **7** and **8**.

Selective 5-*exo-dig* cyclization of pentynoic acids RC≡CCH₂CH₂CO₂H (R = aryl or heteroaryl group) to afford the corresponding (Z)-γ-alkylidene butyrolactones could also be achieved, in water at 50 °C, using the related amphiphilic NCN-pincer Pd(II) complex **8** (2 mol %) (Figure 1) in combination with Et₃N (6 mol %) (yields in the range 38–94% after 1–6 h) [34]. Complex **8** self-assembles in aqueous solution forming bilayered vesicles which were found to be essential for the promotion of the catalytic process (markedly lower yields were obtained when the same reactions were performed in organic solvents or employing the amorphous complex **8** in water).

On the other hand, despite the well-known ability of platinum compounds to promote the cycloisomerization of polyunsaturated organic molecules [35,36], this metal has been much less used than palladium to catalyze the cyclization of alkynoic acids [37]. In this regard, the work published by Alemán, Navarro-Ranninger and co-workers merits highlighting [38]. These authors evaluated the catalytic behaviour of a broad family of anticancer platinum(II) and platinum(IV) amino-complexes (**9a–j** and **9k–o**, respectively; see Figure 2) in the cycloisomerization of 4-pentynoic acid into 5-methylene-dihydrofuran-2-one (5-*exo-dig* cyclization), finding that both the oxidation state of the metal and the stereochemistry of the complex are key factors for the reaction to proceed efficiently. Thus, while the *cis*-platinum(II) derivatives **9a–e** showed a high reactivity in acetone at r.t., their *trans*-platinum(II) counterparts **9f–j** and the Pt(IV) species **9k–o** (regardless of their *cis* or *trans* configuration) turned out to be practically inactive.

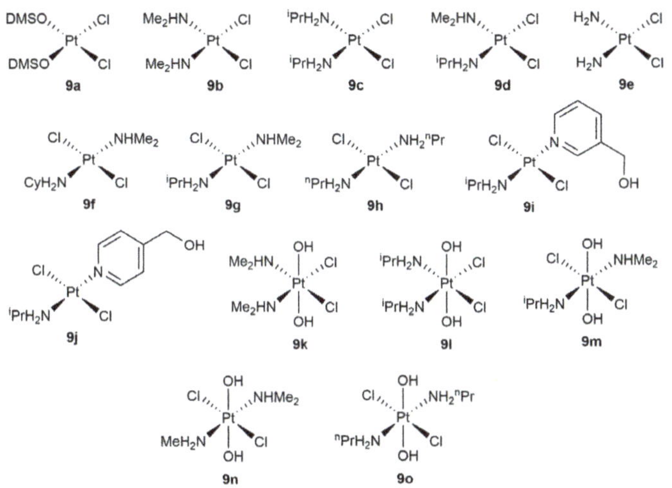

Figure 2. Structure of the platinum complexes **9a–o**.

In addition, they also demonstrated the possibility of using water (or even blood plasma) as solvent. In fact, the scope of the catalytic reaction was explored in water using complexes **9a,b** [38,39]. Thus, as shown in Scheme 5, different 5- and 6-membered ring lactones could be synthesized with these complexes, although mixtures of regioisomers were systematically formed starting from internal alkynes. On the other hand, it is also worth noting that the attempt made to generate a seven-membered ring lactone with this type of catalysts failed.

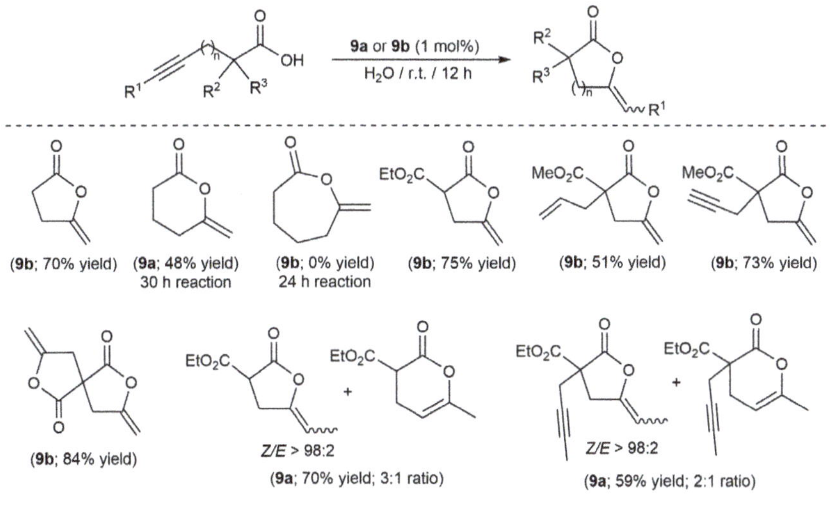

Scheme 5. Cyclization of different alkynoic acids in water using Pt(II) complexes.

Gold compounds are currently recognized as the most effective systems for the electrophilic π-activation of unsaturated carbon–carbon bonds [40,41], and, in 2006, Michelet and co-workers demonstrated for the first time their usefulness in the cycloisomerization of acetylenic acids [42]. The same group was also a pioneer in the use of gold catalysts in aqueous environments. In particular, in 2008, they reported on the tolerance of the heterogeneous system Au_2O_3 towards the presence of water during the cycloisomerization of 2-phenyl-4-pentynoic acid into 5-methylene-3-phenyl-dihydrofuran-2-one (92% yield performing the reaction in MeCN-H_2O (6:1) at r.t. with 2.5 mol % of Au_2O_3 for 3 h; 95% yield when acetonitrile was employed alone under identical reaction conditions) [43]. Since then, a number of gold-based catalysts capable of operating in pure water or in aqueous biphasic mixtures have been described, most involving functionalized *N*-heterocyclic carbenes (NHC) as auxiliary ligands. In this regard, the groups of Michelet, Cadierno and Conejero developed the zwitterionic Au(III)-NHC complex **10b** which proved to be active in the cycloisomerization of a broad range of γ-alkynoic acids under biphasic toluene/water conditions (Scheme 6) [44]. Remarkably, the participation of a silver(I) co-catalyst, usually employed in catalytic gold chemistry to generate vacant coordination sites on the metal through chloride ligands abstraction, was not required. Also of note is the fact that, despite the well-known ability of gold complexes to promote the hydration of alkynes, competitive hydration processes were not observed under the biphasic conditions employed, even during the cycloisomerization of bispropargylic substrates. However, it should be noted that, if pure water is used as solvent, partial hydrolysis of the lactone products takes place, making the use of biphasic conditions more advantageous. In general, the reactions proceeded in air under very mild temperature conditions (r.t.), except with those substrates containing internal C≡C bonds which showed a markedly lower reactivity and required of heating at 80 °C. Concerning the regioselectivity of the process, 5-membered ring enol-lactones were selectively formed when terminal C≡C bonds were involved in the cyclization process. On the other hand, although the 5-*exo-dig* cyclization was also the preferred reaction pathway with internal alkynes, mixtures containing the corresponding 5- and 6-membered ring lactones were in some cases obtained.

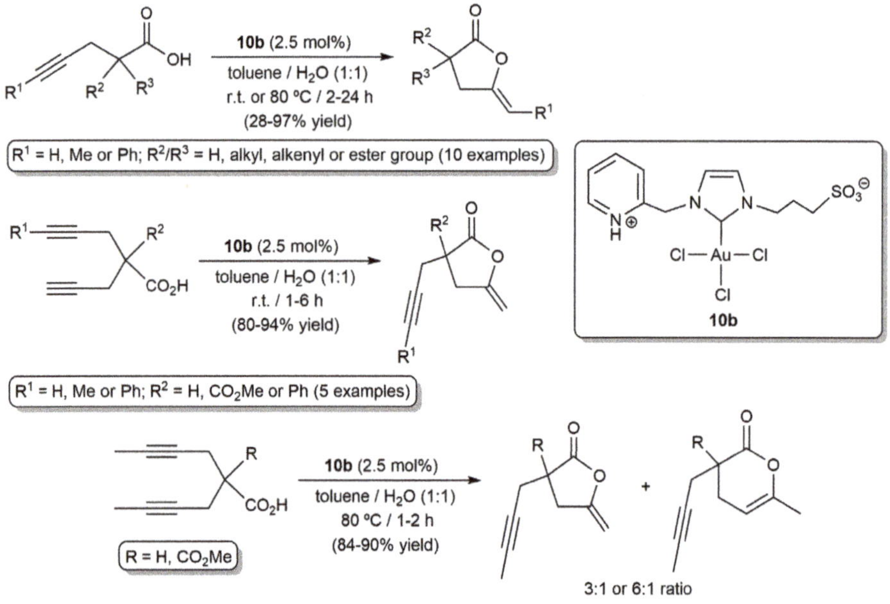

Scheme 6. Cyclosiomerization of different γ-alkynoic acids catalyzed by the Au(III)-*N*-heterocyclic carbenes (NHC) complex **10b**.

This initial work with **10b** was subsequently extended to other Au(III) and Au(I) complexes containing related zwitterionic NHC ligands bearing 3-sulfonatopropyl, and 2-pyridyl, 2-pycolyl or pyridylethyl substituents (**10a,c** and **11a–c** in Figure 3) [45]. All of them proved to be catalytically active, even with Au loadings of only 0.1 mol %, showing in general performances similar to that of **10b**. Interestingly, all these gold-carbenes showed a very high recyclability (up to 10 consecutive runs) by simple phase separation (the gold catalyst remained in the aqueous phase while the lactone product was completely dissolved in the toluene one). In this regard, some differences were observed between the Au(I) and Au(III) species, the recyclability of the latter being much more effective due to their higher stability in the aqueous medium (the Au(I) derivatives **11a–c** undergo with time partial decomposition into catalytically inactive Au(0) nanoparticles; a process also observed with **10a–c** but which takes place much more slowly).

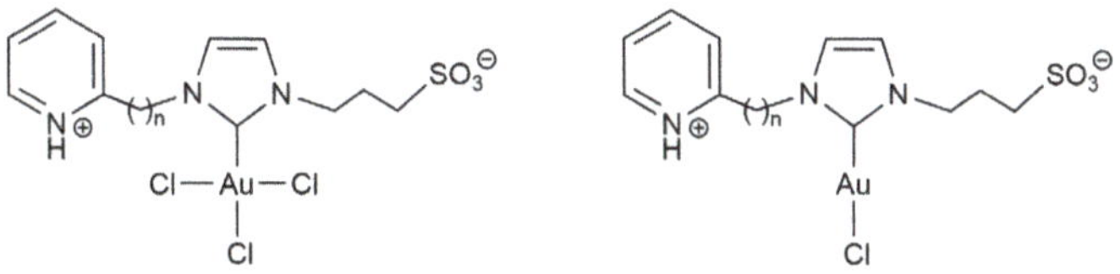

Figure 3. Structure of the zwitterionic Au(III)- and Au(I)-NHC complexes **10,11a–c**.

A good recyclability was also reported by Pleixats and co-workers for the sol-gel immobilized Au(I)-NHC complex **12** (Scheme 7) [46]. This organometallic hybrid silica material proved to be active in the cycloisomerization of different γ-alkynoic acids at room temperature using, as in the previous example, a toluene/water biphasic system and in the absence of silver salts. However, due to the heterogeneous nature of **12**, longer reaction times and a wrist-type shaker stirring (which allows a good mixing of the immiscible layers and the insoluble catalyst) were in this case required to obtain the enol-lactone products in high yields (no reaction was observed using conventional magnetic stirring). Concerning the regioselectivity of the process, five-membered enol-lactones were selectively formed starting from substrates with terminal alkyne units, while a mixture of the 5- and 6-membered ring lactone products was obtained from an internal diyne (Scheme 7).

For its part, the group of Krause described the preparation of the ammonium salt-tagged Au(I)-NHC complexes **13a–f** (Figure 4) and their application in the selective 5-*exo-dig* cyclization of γ-alkynoic acids bearing a terminal alkyne unit in pure water, or in aqueous triethylammonium buffer solution [47]. The catalytic reactions proceeded cleanly at r.t., in short time spans (0.5–6 h), employing 2.5 mol % of these complexes, with yields higher in general when the buffer solution was employed as the reaction medium (partial decomposition of the gold complexes was observed in pure water). Once again, no Ag(I) co-catalysts were needed and the catalysts could be reused after extraction of the lactone product from the aqueous solution with diethyl ether (up to 5 times).

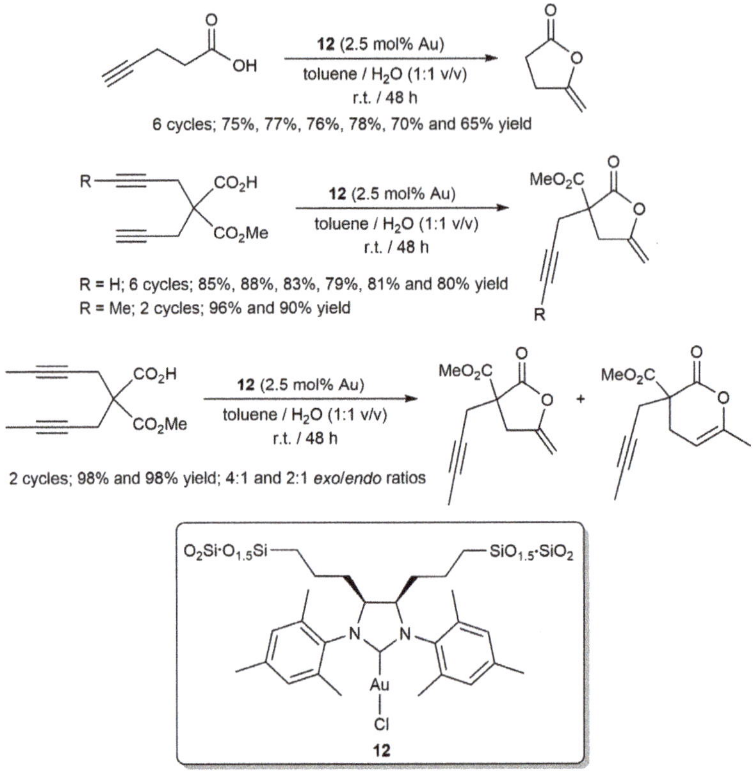

Scheme 7. Cyclosiomerization of γ-alkynoic acids using a sol-gel immobilized Au(I)-NHC complex.

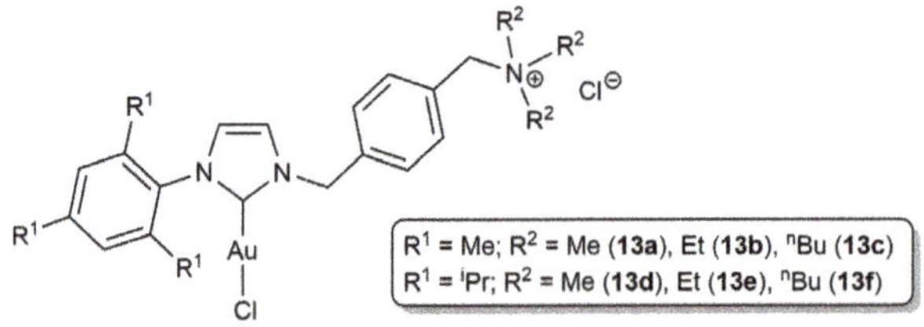

Figure 4. Structure of the water-soluble gold(I)-NHC complexes **13a–f**.

Interestingly, starting from the internal alkynoic acid **14**, a synthetic route to obtain the furanocoumarin-functionalized lactone **15**, an epimer of the natural product clausemarine A, could be developed by Krause and co-workers [47]. As shown in Scheme 8, formation of the lactone ring was successfully achieved by cycloisomerization of **14** with 1 mol % of **13b** in an aqueous triethylammonium buffer solution containing THF (tetrahydrofuran) as co-solvent.

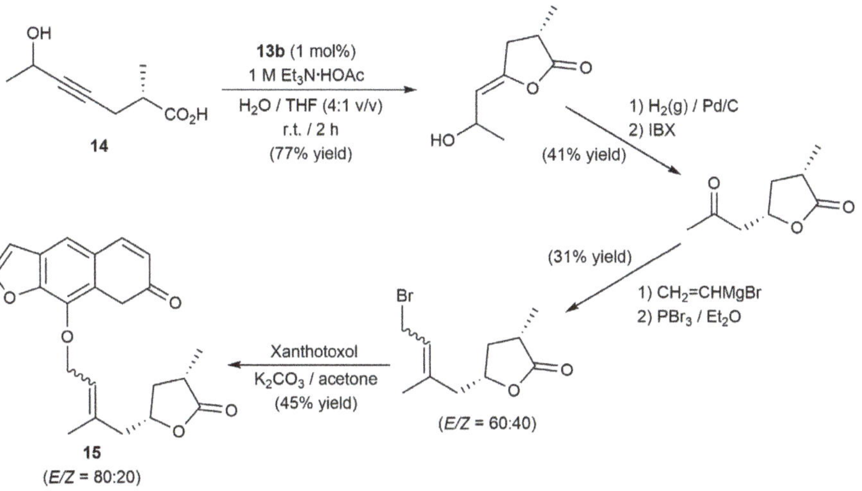

Scheme 8. Synthesis of 2-*epi*-clausemarine A.

Besides the Au-NHC complexes commented above, the PTA-derived iminophosphorane-Au(I) derivative **16** (Figure 5) proved to be also an active and selective catalyst for the 5-*exo-dig* cyclization of 4-pentynoic acid in water (89% yield after 30 min at r.t. with 1 mol % of **16**) [48]. However, this particular catalyst showed a higher reactivity in the eutectic mixture *1ChCl/2Urea* (ChCl = choline chloride; 99% yield after 15 min under identical reaction conditions) and, consequently, its scope was explored only in this alternative and biorenewable reaction medium.

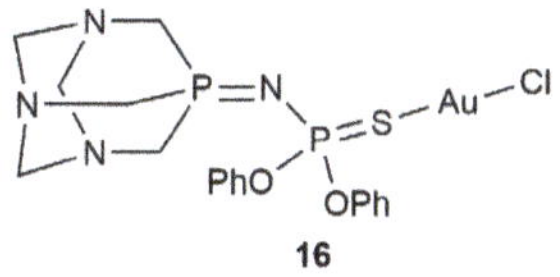

Figure 5. Structure of the iminophosphorane-gold(I) complex **16**.

In addition to palladium, platinum and gold, cheaper copper catalysts have also been used in the cycloisomerization of alkynoic acids in aqueous media. Thus, while studying the CuBr-catalyzed cycloaddition of alkynes with azides in ^{t}BuOH/H$_2$O mixtures, Mindt and Schibli observed the formation of enol-lactone by-products when employing γ-alkynoic acids as substrates, so they decided to explore separately the cyclization of these compounds [49]. Their results are shown in Scheme 9. Performing the reactions with 10 mol % of CuBr at r.t., different γ-alkynoic acids containing both terminal and internal C≡C units could be efficiently transformed into the corresponding γ-alkylidene butyrolactones. However, in some cases, γ-keto acids were selectively obtained due to the rapid hydrolysis of the lactone products under the aqueous conditions employed. On the other hand, the attempts made to extend the scope of this aqueous protocol to six- and seven-membered ring enol-lactones by cyclization of 5-hexynoic acids and 6-heptynoic acids, respectively, failed. It should be noted, however, that the former could be cyclized by carrying out the reactions in acetonitrile instead of the ^{t}BuOH/H$_2$O mixture.

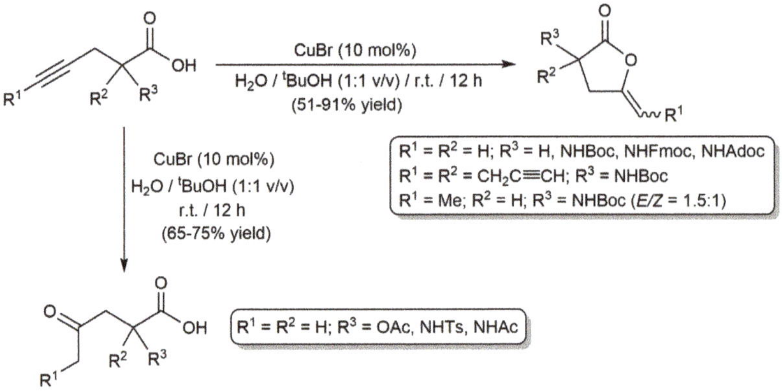

Scheme 9. CuBr-catalyzed transformations of different γ-alkynoic acids in aqueous medium.

More recently, CuBr was also employed to promote the cyclization of a variety of β-hydroxy-γ-alkynoic acids **17** [50]. As shown in Scheme 10, the reactions proceeded cleanly in pure water under microwave (MW) irradiation at 100 °C, leading to the regio- and stereoselective formation of the five membered ring Z-enol-lactones **18** without observing hydrolysis products. In addition, the beneficial effect of water was evidenced by the authors, who obtained markedly lower yields when performing the same reactions in classical organic solvents such as THF or acetonitrile. The same must be said of the use of MW irradiation, since lower yields were also observed when conventional oil-bath heating or ultrasound irradiation was employed. The process was quite general, tolerating a broad range of substitution patterns on the substrate skeleton. However, it should be noted that when an alkynoic acid featuring a hydrogen atom and a phenyl group, as the R^3 and R^4 substituents, was employed in this reaction, the spontaneous dehydration of **18** to form a 5-alkylidene-furan-2-one product took place.

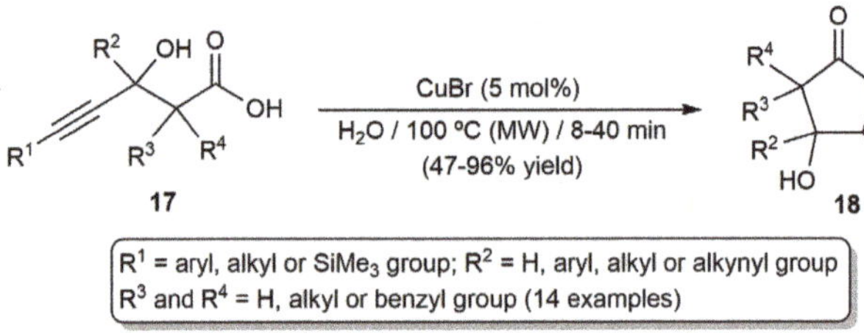

Scheme 10. Cu(I)-catalyzed cyclization of β-hydroxy-γ-alkynoic acids in water under microwave (MW) irradiation.

For its part, the group of Jiao described the cyclization of the propargylic Meldrum's acids **19** into the (Z)-γ-alkylidene butyrolactones **20**, in a basic aqueous environment, through the combined use of Cu(OAc)$_2$ and FeCl$_3$ (Scheme 11) [51]. A clear cooperative effect of the two metals was observed, the yields decreasing drastically when Cu(OAc)$_2$ or FeCl$_3$ was used as the sole catalyst. The process most probably involves the in situ generation of a γ-alkynoic acid intermediate through a hydrolysis/decarboxylation sequence. On the other hand, although the mechanism could not be unambiguously established, it was assumed that the Cu^{2+} ion is the one responsible for the activation of the alkyne unit during the final cycloisomerization step, with Fe^{3+} probably facilitating the intramolecular nucleophilic addition of the carboxylate on the C≡C bond by coordination to the carbonyl group.

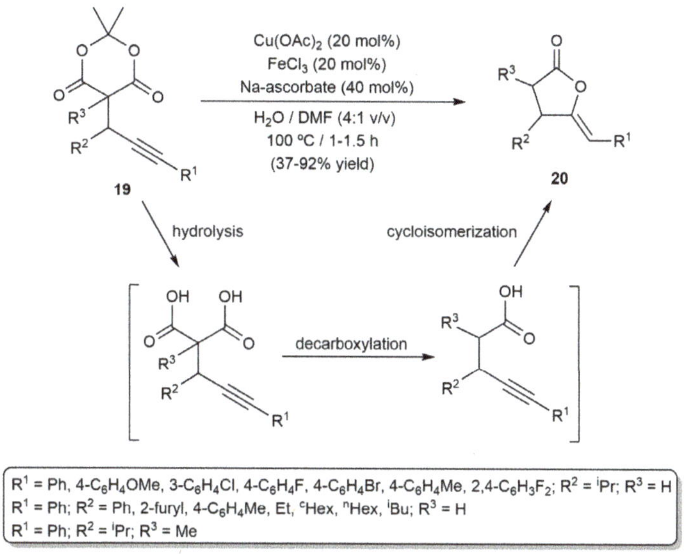

Scheme 11. Cu/Fe-cocatalyzed synthesis of lactones from propargylic Meldrum's acids.

Further studies by the same group showed that the same catalytic reactions can be more conveniently carried out employing AgNO$_3$ (5 mol %) instead of the Cu/Fe system [52]. The reactions, performed at 100 °C in a H$_2$O/DMF (N,N-dimethylformamide) mixture under air, afforded regio- and stereoselectively the (Z)-γ-alkylidene butyrolactone products **20** in 45–87% isolated yields after ca. 1 h. Remarkably, no co-catalysts or bases were in this case needed. On the other hand, transformation of the related Meldrum´s acids **21** into lactones **22**, containing an all-carbon quaternary center at the C-4 position, was also described by Ahmar and Fillion using catalytic amounts of Ag$_2$CO$_3$, and mixtures THF/H$_2$O or toluene/H$_2$O as the reaction media (Scheme 12) [53]. Again, the 5-*exo-dig* cyclization products were exclusively formed and the reactions proceeded cleanly in the absence of base. However, we must note that mixtures of *E/Z* isomers were in most cases obtained starting from those substrates containing an internal C≡C bond.

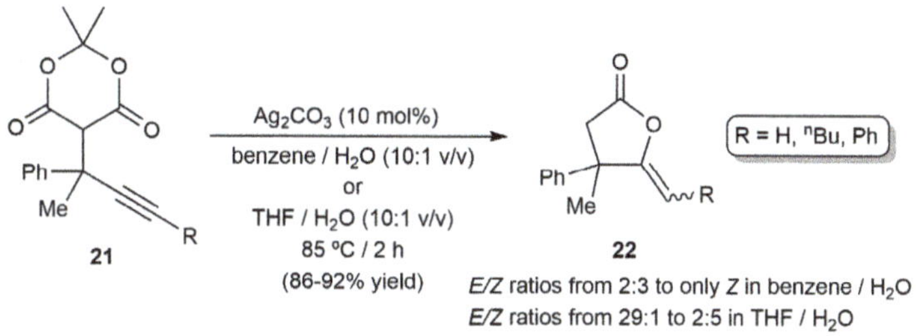

Scheme 12. Silver-catalyzed cyclization of the propargylic Meldrum's acids.

2.2. Tandem and Cascade Processes Involving the Cycloisomerization of an Alkynoic Acid

The aminolysis of lactones is a common transformation in organic synthesis which allows their direct conversion into linear amides [54,55]. The combination of this reaction with the cycloisomerization of alkynoic acids has been extensively studied in the last years giving access to a huge number of nitrogen-containing heterocyclic compounds, through different cascade processes, by appropriate selection of the functionalized amine partner [56–59]. In this context, Liu and co-workers developed an efficient and broad scope gold-catalyzed reaction for the synthesis of fused polycyclic indoles **23** in water, starting from 2-(1*H*-indol-1-yl)alkylamines and alkynoic acids (Scheme 13) [60].

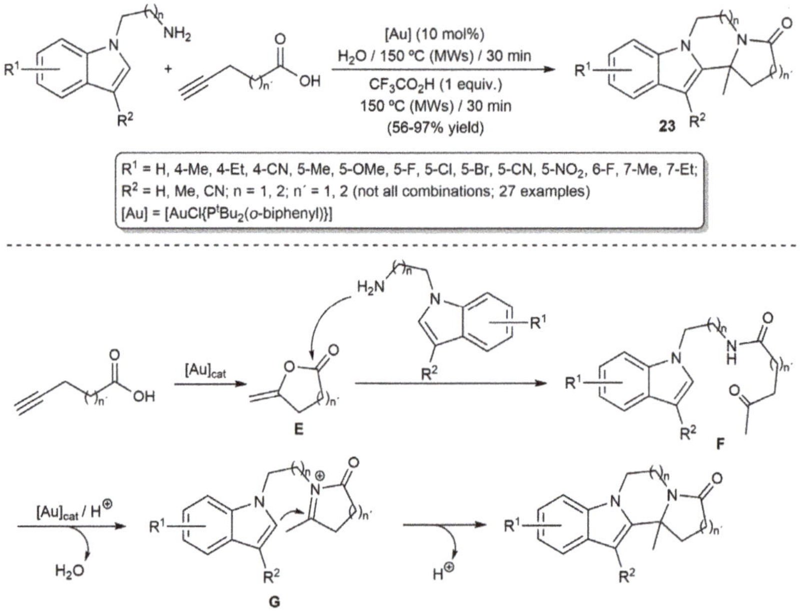
[Au] (10 mol%)
H_2O / 150 °C (MWs) / 30 min
CF_3CO_2H (1 equiv.)
150 °C (MWs) / 30 min
(56-97% yield)
23
R^1 = H, 4-Me, 4-Et, 4-CN, 5-Me, 5-OMe, 5-F, 5-Cl, 5-Br, 5-CN, 5-NO_2, 6-F, 7-Me, 7-Et;
R^2 = H, Me, CN; n = 1, 2; n′ = 1, 2 (not all combinations; 27 examples)
[Au] = [AuCl{P^tBu_2(o-biphenyl)}]
$[Au]_{cat}$
E
F
$[Au]_{cat}$ / $H^{\oplus}$
H_2O
G
$H^{\oplus}$

Scheme 13. Gold-catalyzed synthesis of fused polycyclic indoles in water.

The process involves the aminolysis of the in situ formed enol-lactones **E** to generate the linear keto-amides **F**, which subsequently evolve into **G** via a gold-catalyzed *N*-acyliminium ion formation/cyclization. Final intramolecular nucleophilic attack of the C-2 carbon of the indolic unit to the iminium carbon yields the products **23**. To facilitate the cyclization step leading to **G**, the addition trifluoroactic acid was in some cases required. As shown in Scheme 13, the reactions proceeded in short times under microwave irradiation, tolerating the presence of different functional groups.

The same group also described the coupling of 4-pentynoic acid with *o*-aminophenylmethanol in water catalyzed by the cationic gold(I) complex [Au{P^tBu_2(*o*-biphenyl)}(MeCN)][SbF_6] (Scheme 14) [61]. The reaction afforded the pyrrolo[2,1-*b*]benzo[*d*][1,3]oxazin-1-one **24** in 50% yield. However, we must note that this value was much lower to that obtained in THF under identical reaction conditions (91% yield), so it is not surprising that the generality of the process was studied in the latter solvent. As in the precedent case, the reaction is initiated by the cycloisomerization of the alkynoic acid, subsequent aminolysis of the enol lactone intermediate, and final cyclization of the resulting keto amide catalyzed by the gold complex.

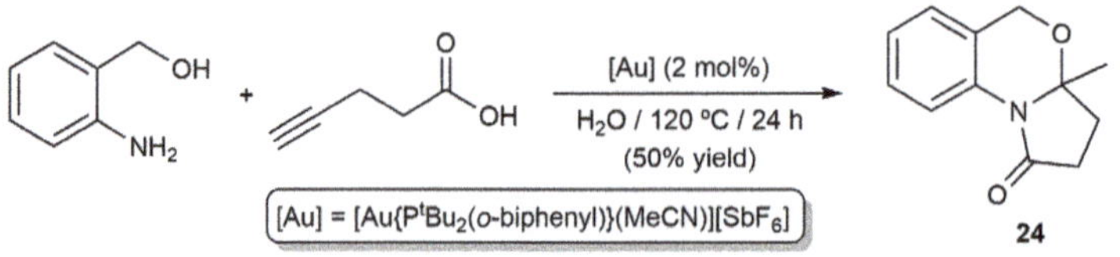
[Au] (2 mol%)
H_2O / 120 °C / 24 h
(50% yield)
[Au] = [Au{P^tBu_2(o-biphenyl)}(MeCN)][SbF_6]
24

Scheme 14. Gold-catalyzed synthesis of the pyrrolo[2,1-*b*]benzo[*d*][1,3]oxazin-1-one **24** in water.

In another vein, the group of Jiang developed an aqueous protocol for the one-pot synthesis of phthalides **26** from terminal alkynes and *o*-iodobenzoic acid (Scheme 15) [62]. The process, which involves the initial Sonogashira coupling of the substrates and subsequent stereoselective 5-*exo-dig* cyclization of the resulting ethynyl-benzoic acid intermediates **25**, was catalyzed by palladium immobilized on carbon nanotubes (CNTs) in a DMF:H_2O mixture. Unlike previous examples of this tandem reaction, formation of isocoumarin by-products (6-*endo-dig* cyclization) was in this case not observed, and no additives such as phosphine ligands or CuI were needed.

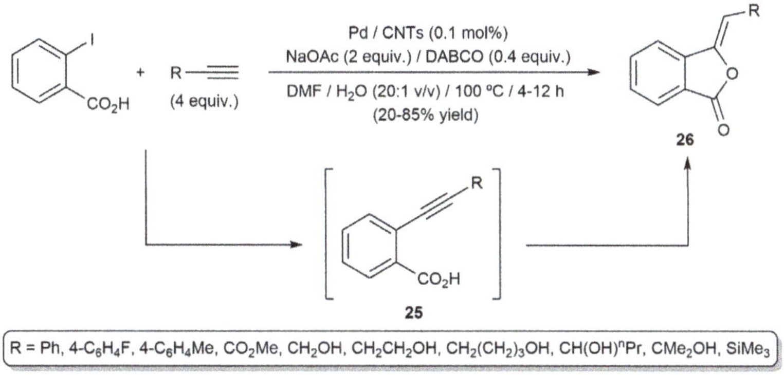

Scheme 15. Catalytic synthesis of phthalides from terminal alkynes and *o*-iodobenzoic acid.

Taking advantage of the ability of the dinuclear iminophosphorane-palladium(II) complex **4** to promote the fast and selective transformation of bispropargylic carboxylic acids **5** into enol-lactones **6** (Scheme 4), García-Álvarez and co-workers could set up an unprecedented one-pot tandem process combining the aforementioned reaction with a copper-catalyzed 1,3-dipolar cycloaddition of azides with the terminal alkyne arm of the enol-lactone products (Scheme 16) [30]. To promote the 1,3-dipolar cycloaddition step the polymeric Cu(I) catalyst $[Cu\{\mu^2\text{-}N,S\text{-}(PTA)\text{=}NP(\text{=}S)(OEt)_2\}]_x[SbF_6]_x$ (**27**) containing the same hydrophilic iminophosphorane ligand [63], in combination with the 2,6-lutidine base, was employed. The tandem process proceeded in pure water under mild conditions, affording the bicyclic triazol-enol-lactones **28** in excellent yields after a simple extraction with diethyl ether (no chromatographic purification was needed).

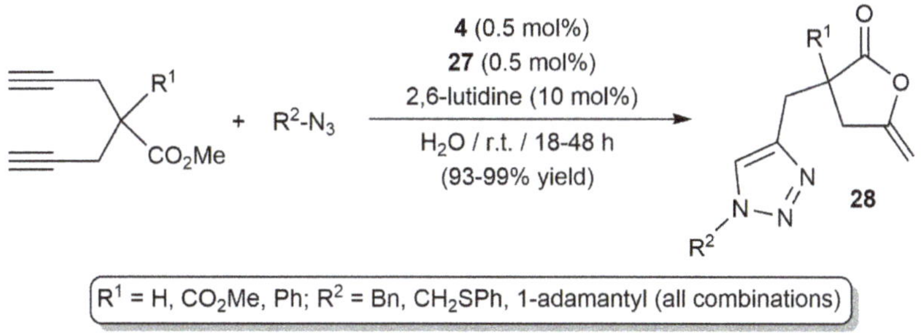

Scheme 16. Access to bicyclic triazol-enol-lactones from of bispropargylic carboxylic acids and azides.

On the other hand, based on the capability of copper salts to promote the conjugate alkynylation of electron-deficient olefins, the (Z)-γ-alkylidene butyrolactones **30** could be synthesized in a one-pot manner starting from the corresponding terminal alkynes and the 5-alkylidene-Meldrun's acids **29**, via in situ formation of the corresponding propargylic Meldrun's acids (Scheme 17). The process was promoted by the $Cu(OAc)_2/FeCl_3$ combination discussed above for the cyclization of Meldrun´s acids **19** (Scheme 11) [51].

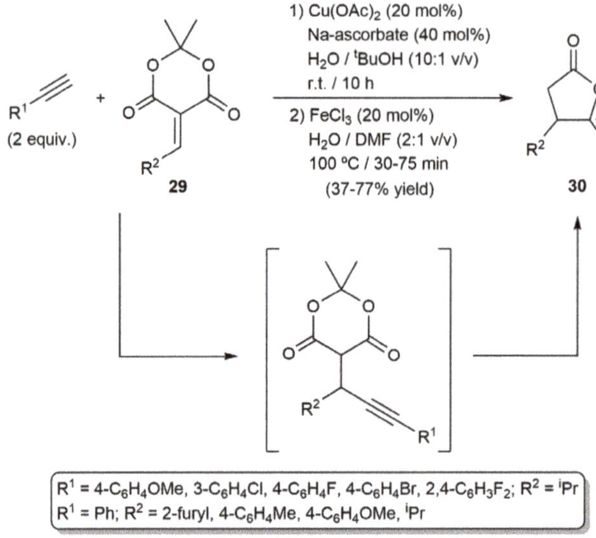

Scheme 17. Cu/Fe-cocatalyzed synthesis of lactones from alkynes and 5-alkylidene-Meldrum's acids.

Finally, the combination of metal catalysis and biocatalysis has emerged in recent years as a powerful tool for developing new synthetic methodologies merging the advantages of both disciplines in terms of reaction scope and selectivity [64–66]. In this context, García-Álvarez, González-Sabín and co-workers described very recently the one-pot conversion of 4-pentynoic acid into enantiopure γ-hydroxyvaleric acid in aqueous medium, through the combined use of $KAuCl_4$ and ketoreductases (KREDs) (Scheme 18) [67]. The process involves the initial gold-catalyzed cycloisomerization of the substrate, concomitant hydrolysis of the lactone to form levulinic acid, and final bioreduction of the keto group of the latter. Remarkably, no isolation or purification steps were required, the reaction medium coming from the metal-catalyzed reaction being directly employed in the enzymatic step. Also of note is the fact that, just by selecting the adequate KRED, both enantiomers of the γ-hydroxyvaleric acid could be obtained with excellent *ee* values.

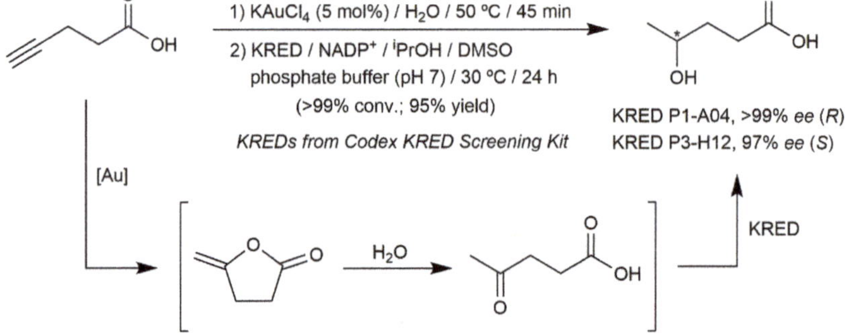

Scheme 18. Chemoenzymatic one-pot conversion of 4-pentynoic acid into enantiopure γ-hydroxyvaleric acid.

3. Intermolecular Processes

3.1. Catalytic Addition of Carboxylic Acids to Terminal and Internal Alkynes

Although the intermolecular hydro-oxycarbonylation of alkynes has been widely studied in organic media [1–6], to date there are very few examples of this catalytic transformation described in water. In this regard, Cadierno, Gimeno and co-workers evaluated in 2011 the catalytic potential of a diverse family of ruthenium(IV) complexes, of general composition *trans*-[RuCl$_2$(η^3:η^3-C$_{10}$H$_{16}$)(L)] (**31a–s** in Figure 6), for the hydro-oxycarbonylation of terminal alkynes in aqueous medium [68].

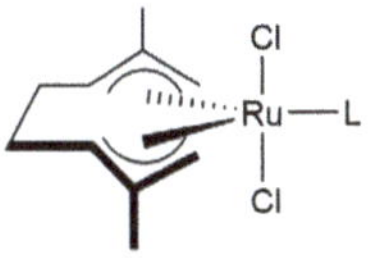

L = PPh$_3$ (**31a**), TPPMS (**31b**), P(p-Tol)$_3$ (**31c**), PPh$_2$Me (**31d**), PPhMe$_2$ (**31e**), PMe$_3$ (**31f**), P^{i}Pr$_3$ (**31g**), PBn$_3$ (**31h**), P(OMe)$_3$ (**31i**), P(OEt)$_3$ (**31j**), P(O^{i}Pr)$_3$ (**31k**), P(OPh)$_3$ (**31l**), NCMe (**31m**), NCPh (**31n**), NH$_2$Ph (**31o**), py (**31p**), CO (**31q**), CNBn (**31r**), CNCy (**31s**)

Figure 6. Structure of the bis(allyl)-ruthenium(IV) complexes **31a–s**.

All these compounds were found to be active catalysts in the addition of benzoic acid to 1-hexyne in pure water, providing the corresponding enol esters ^{n}BuC(OBz)=CH$_2$ (Markovnikov addition product) and (E/Z)-^{n}BuCH=CH(OBz) (*anti*-Markovnikov addition products) in moderate to good yields after 3–24 h of heating at 60 °C (with a Ru loading of 2 mol %). A high selectivity towards the Markovnikov addition product was in general observed, except in the case of catalysts **31m–p** containing a labile amine or nitrile ligand which generated preferentially (E/Z)-^{n}BuCH=CH(OBz), albeit only in moderate yields. The best results in terms of activity and regioselectivity were obtained with [RuCl$_2$(η^3:η^3-C$_{10}$H$_{16}$)(PPh$_3$)] (**31a**), which was able to generate the enol ester ^{n}BuC(OBz)=CH$_2$ in 96% yield (by gas chromatography) after only 3 h of heating. However, from the data obtained, no relationships between the steric and/or electronic nature of the auxiliary ligand L and the catalytic activity observed could be drawn. On the other hand, it must be also highlighted that, with the exception of [RuCl$_2$(η^3:η^3-C$_{10}$H$_{16}$)(TPPMS)] (**31b**; TPPMS = 3-(diphenylphosphino)benzenesulfonate sodium salt) and [RuCl$_2$(η^3:η^3-C$_{10}$H$_{16}$){P(OR)$_3$}] (R = Me (**31i**), Et (**31j**), ^{i}Pr (**31k**)), these Ru(IV) complexes are completely insoluble in water. Accordingly, in most of the reactions the catalyst remained dissolved in the organic phase, forming an emulsion with water under stirring. In other words, the reactions proceeded under the so-called "on-water" conditions [69,70]. In addition, when the same reactions were carried out under homogeneous conditions using organic solvents, the results were much worse in terms of yields (the regioselectivity remained unaffected), pointing out the marked positive effect of water in the process. Although to a lesser extent, the beneficial effect of water in the catalytic addition of carboxylic acids to terminal alkynes promoted by tethered arene-ruthenium(II) complexes was also observed by Demonceau and co-workers, who obtained higher yields when performing the reactions in moist vs. dry toluene [71].

The most active catalyst of the series, i.e., [RuCl$_2$(η^3:η^3-C$_{10}$H$_{16}$)(PPh$_3$)] (**31a**), showed also a wide scope allowing the preparation of a large variety of enol esters **32** by selective Markovnikov addition of different carboxylic acids to both aliphatic and aromatic terminal alkynes, as well as to 1,3-enynes (Scheme 19) [68,72,73]. As a general trend, the reactions proceeded faster, and with higher yields, when aliphatic terminal alkynes were employed as substrates. Also notable are the high functional group compatibility showed by complex **31a** and the absence of competing processes of hydration of the alkynes.

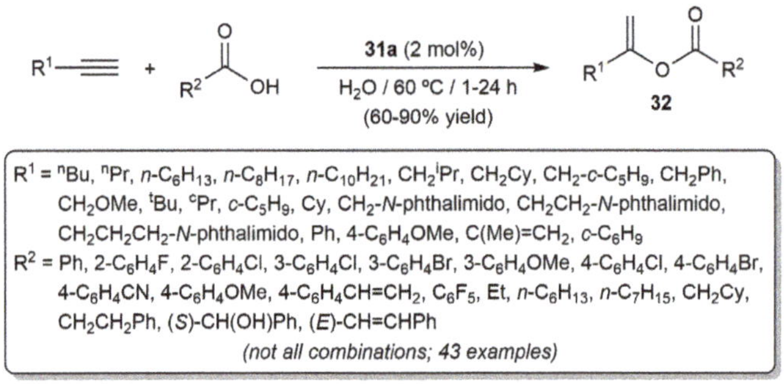

$$R^1 \text{—≡} + R^2\text{C(O)OH} \xrightarrow[\substack{H_2O\,/\,60\,°C\,/\,1\text{-}24\,h \\ (60\text{-}90\%\ \text{yield})}]{\textbf{31a}\ (2\ \text{mol\%})} \textbf{32}$$

R^1 = ^{n}Bu, ^{n}Pr, n-C$_6$H$_{13}$, n-C$_8$H$_{17}$, n-C$_{10}$H$_{21}$, CH$_2$^{i}Pr, CH$_2$Cy, CH$_2$-c-C$_5$H$_9$, CH$_2$Ph,
CH$_2$OMe, ^{t}Bu, ^{c}Pr, c-C$_5$H$_9$, Cy, CH$_2$-N-phthalimido, CH$_2$CH$_2$-N-phthalimido,
CH$_2$CH$_2$CH$_2$-N-phthalimido, Ph, 4-C$_6$H$_4$OMe, C(Me)=CH$_2$, c-C$_6$H$_9$
R^2 = Ph, 2-C$_6$H$_4$F, 2-C$_6$H$_4$Cl, 3-C$_6$H$_4$Cl, 3-C$_6$H$_4$Br, 3-C$_6$H$_4$OMe, 4-C$_6$H$_4$Cl, 4-C$_6$H$_4$Br,
4-C$_6$H$_4$CN, 4-C$_6$H$_4$OMe, 4-C$_6$H$_4$CH=CH$_2$, C$_6$F$_5$, Et, n-C$_6$H$_{13}$, n-C$_7$H$_{15}$, CH$_2$Cy,
CH$_2$CH$_2$Ph, (S)-CH(OH)Ph, (E)-CH=CHPh
(not all combinations; 43 examples)

Scheme 19. Ruthenium(IV)-catalyzed Markovnikov addition of carboxylic acids to terminal alkynes.

Further evidence of the versatility of [RuCl$_2$(η^3:η^3-C$_{10}$H$_{16}$)(PPh$_3$)] (**31a**) was gained in the reactions of the terminal diynes **33** with benzoic acid. Thus, as shown in Scheme 20, the corresponding enynes **34** or the diesters **35** could be selectively synthesized with **31a** just by adjusting the diyne/benzoic acid ratio employed [68,72,73].

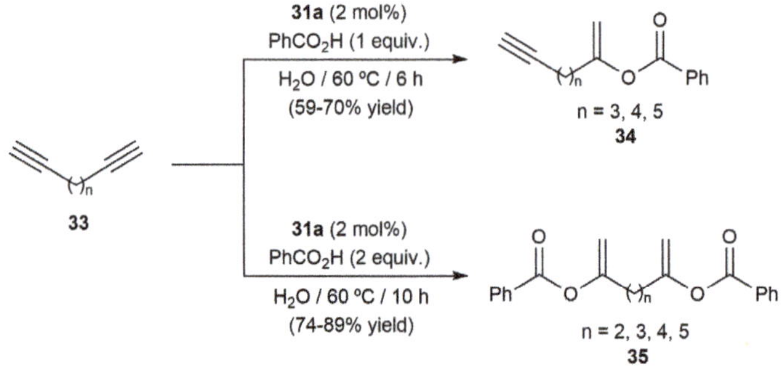

Scheme 20. Ruthenium(IV)-catalyzed Markovnikov addition of benzoic acid to terminal diynes.

The only limitation encountered with the ruthenium catalyst **31a** is that it is completely inactive with internal alkynes; substrates which, as commented in the introduction of this article, show a much lower reactivity in these intermolecular hydrocarboxylation reactions. For this particular class of alkynes, Cadierno, García-Garrido and co-workers developed very recently a protocol in water employing the catalytic system [AuCl(PPh$_3$)]/AgOAc [74]. Thus, as depicted in Scheme 21, a broad range of trisubstituted enol esters **36** could be synthetized in good yields by addition of different aromatic, aliphatic, heterocyclic and α,β-unsaturated carboxylic acids to internal alkynes symmetrically substituted with aliphatic groups. Concerning the use of aromatic alkynes, they showed only a residual reactivity towards benzoic acid and, when an aliphatic carboxylic acid was employed, a higher temperature and catalyst loading were required to obtain the corresponding products in moderate yields. Interestingly, compounds **36** were obtained in a stereoselective manner (only Z isomers) as the result of the exclusive *anti* addition of the O-H bond of the carboxylic acid to the alkyne.

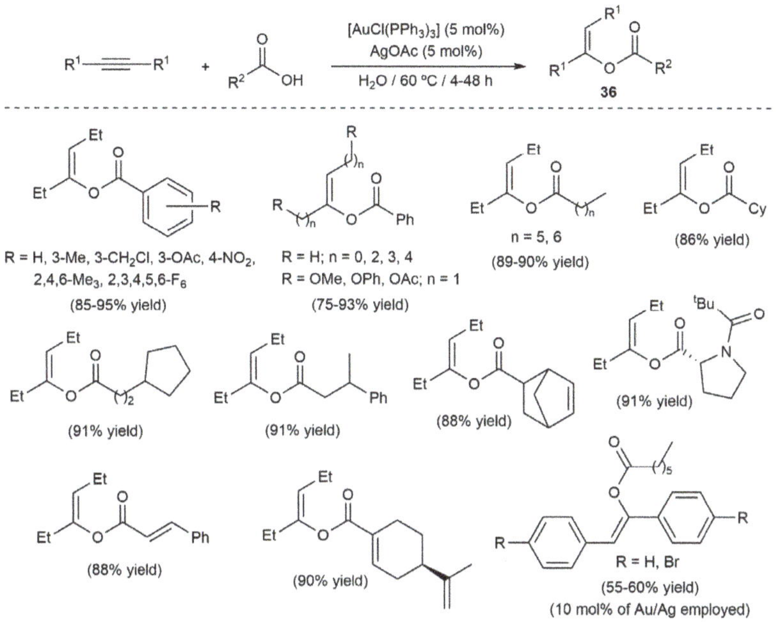

Scheme 21. Gold-catalyzed addition of carboxylic acids to symmetrically substituted internal alkynes.

The reactivity of some unsymmetrically substituted internal alkynes towards benzoic acid in water was also explored using the catalytic system [AuCl(PPh$_3$)]/AgOAc, the reations leading in most of the cases to mixtures of regioisomers derived from the attack of acid to both carbon atoms of the alkyne (complete Z-selectivity was again observed for both regioisomers). Only when the activated internal alkynes **37** and **38** were employed as substrates a complete regioselectivity to enol esters **39** and **40**, respectively, was observed (Scheme 22) [74].

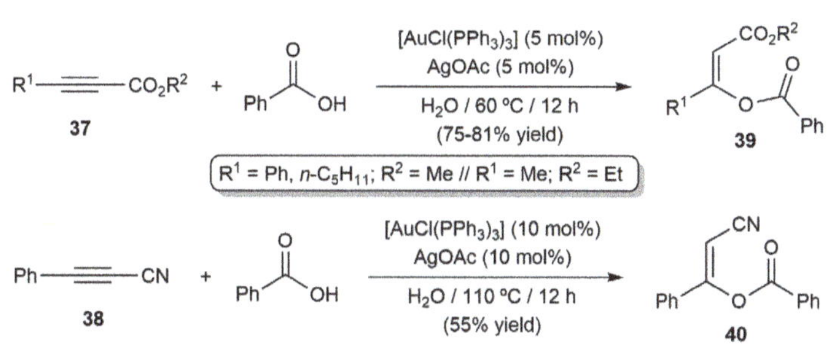

Scheme 22. Regio- and stereoselective addition of benzoic acid to unsymmetrically substituted alkynes.

3.2. Catalytic Addition of Carboxylic Acids to Terminal Propargylic Alcohols

β-Oxo esters are useful intermediates in organic chemistry. For example, they have been employed in the synthesis of different pharmaceuticals [75–77], and can be easily transformed into the corresponding α-hydroxy ketones, which are structural units present in a large variety of biologically active molecules [78,79]. Among the different synthetic approaches to β-oxo esters [80], the ruthenium-catalyzed addition of carboxylic acids to terminal propargylic alcohols has emerged as one of the most straightforward and atom-economical routes (Scheme 23). The process involves the Markovnikov attack of the carboxylic acid to an initially formed π-alkyne-ruthenium complex **H**, which is followed by the intramolecular transesterification of the resulting intermediate **I** to form an alkenyl derivative **J**. The final protonolysis of **J** releases the β-oxo ester product and regenerates the catalytically active ruthenium species.

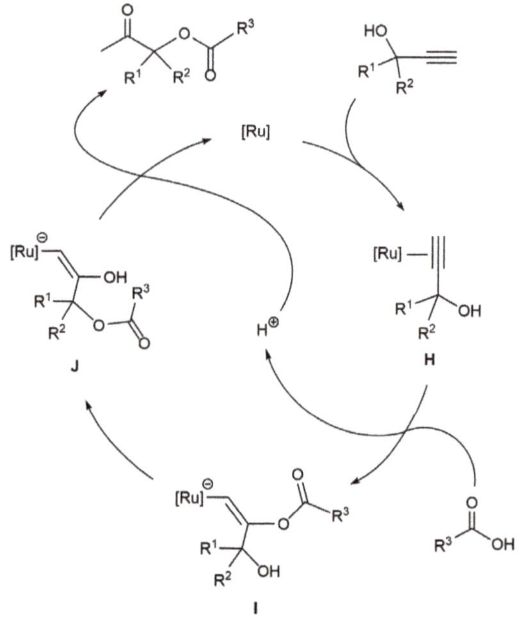
[Ru]
[Ru]
OH
J
[Ru]
OH
H
[Ru]
OH
I
OH

Scheme 23. Mechanism of the Ru-catalyzed β-oxo esters formation reactions.

Since the pioneering work by Mitsudo and Watanabe in 1987 [81], a number of ruthenium catalysts for this transformation have been developed [80,82]. Among them, the bis(allyl)-ruthenium(IV) complex $[RuCl_2(\eta^3:\eta^3\text{-}C_{10}H_{16})(PPh_3)]$ (**31a**) (Figure 6) proved to be active in water [68]. Thus, as shown in Scheme 24, starting from different propargylic alcohols and benzoic acid, a family of β-oxo esters **41** could be synthetized in moderate to good yields employing identical experimental conditions to those applied in the preparation of the enol esters **32** (Scheme 19). Although the scope of the reaction concerning the carboxylic acid partner was not studied in much detail, the authors showed that it is not restricted to benzoic acid since the addition of 2-chlorobenzoic acid, pentafluorobenzoic acid, heptanoic acid and 3-cyclopentylpropionic acid to 1-phenyl-2-propyn-1-ol also afforded the corresponding β-oxo esters in 47–90% yield.

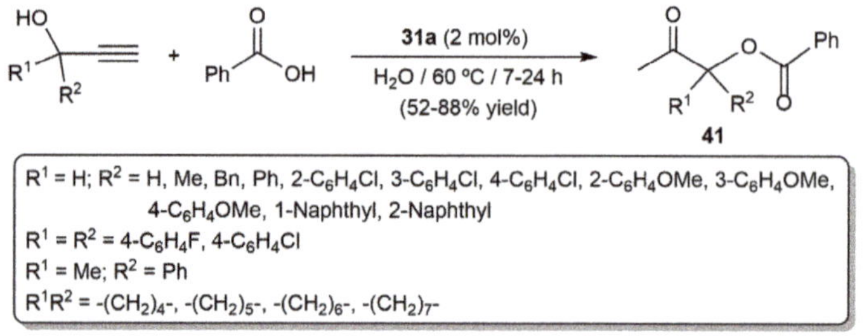
31a (2 mol%)
H2O / 60 °C / 7-24 h
(52-88% yield)
41
R1 = H; R2 = H, Me, Bn, Ph, 2-C6H4Cl, 3-C6H4Cl, 4-C6H4Cl, 2-C6H4OMe, 3-C6H4OMe, 4-C6H4OMe, 1-Naphthyl, 2-Naphthyl
R1 = R2 = 4-C6H4F, 4-C6H4Cl
R1 = Me; R2 = Ph
R1R2 = -(CH2)4-, -(CH2)5-, -(CH2)6-, -(CH2)7-

Scheme 24. Synthesis of β-oxo esters in water catalyzed by the ruthenium(IV) complex **31a**.

In an independent work, Cadierno, Gimeno and co-workers also explored the catalytic potential of a series of arene-ruthenium(II) complexes containing different water-soluble phosphine ligands (**42–45a–d** in Figure 7) [83]. All of them proved to the active in the addition of benzoic acid to 1-phenyl-2-propyn-1-ol in pure water, leading to the desired β-oxo ester, i.e., 1-phenyl-2-oxopropyl benzoate, as the major reaction product. Among them, best results in terms of activity and selectivity were obtained with the benzene derivative [RuCl$_2$(η^6-C$_6$H$_6$)(TPPMS)] (**45a**) (88% yield of 1-phenyl-2-oxopropyl benzoate after 3 h of heating at 100 °C using a ruthenium loading of 2 mol %). Concerning the scope of this complex, a variety of secondary propargylic alcohols, as well as prop-2-yn-1-ol, could be efficiently transformed. In contrast, tertiary propargylic alcohols resulted to be more challenging substrates and only those featuring low sterically-demanding substituents led to high conversions. On the other hand, **45a** was operative with a large variety of aromatic carboxylic acids, bearing different functional group such as halide, alkoxy, ketone or sulfonamide. The use of heteroaromatic acids, with tetrahydrofuran, pyrrole, thiophene, indole or 2-oxo-2*H*-chromene fragments, as well as aliphatic acids was also tolerated.

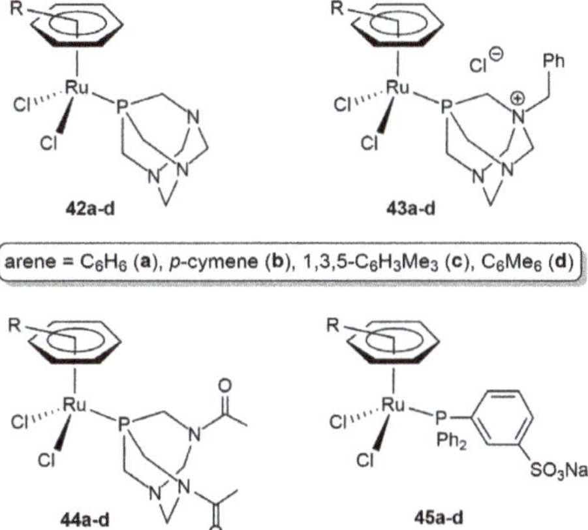

Figure 7. Structure of the water-soluble arene-ruthenium(II) complexes **42–45a–d**.

4. Conclusions

Great attention is currently devoted to synthetic organic chemistry in water and research, particularly in the field of aqueous catalysis, is increasing exponentially since water is the most environmentally benign substitute for the volatile and toxic organic solvents commonly used in laboratories and industries. In this contribution, we have summarized the developments achieved in the field of metal-catalyzed additions of carboxylic acids to alkynes in aqueous media. Such processes now represent powerful tools for the construction of synthetically useful lactones, enol esters and β-oxo esters in an atom-economical manner. Throughout this review article, we have presented different catalytic systems based on Pd, Pt, Au, Cu and Ru, capable of promoting selectively these reactions in aqueous environments without observing the competing hydration of the alkyne substrates. The vast majority of the works discussed herein have been published during the last ten years, demonstrating clearly the current interest in this research field, which obviously remains open, with many opportunities for new discoveries.

Acknowledgments: The authors acknowledge the Spanish MINECO (projects CTQ2013-40591-P and CTQ2016-75986-P) and the Gobierno del Principado de Asturias (project GRUPIN14-006) for financial support. Javier Francos thanks The Ministry of Economy and Competitiveness (MINECO) and European Social Fund (ESF) for the award of a Juan de la Cierva contract.

Author Contributions: Both authors analyzed the data and jointly wrote the paper.

Conflicts of Interest: The authors declare no conflict of interest.

References

1. Alonso, F.; Beletskaya, I.P.; Yus, M. Transition-metal-catalyzed addition of heteroatom-hydrogen bonds to alkynes. *Chem. Rev.* **2004**, *104*, 3079–3159. [CrossRef] [PubMed]
2. Beller, M.; Seayad, J.; Tillack, A.; Jiao, H. Catalytic Markovnikov and anti-Markovnikov functionalization of alkenes and alkynes: Recent developments and trends. *Angew. Chem. Int. Ed.* **2004**, *43*, 3368–3398. [CrossRef] [PubMed]
3. Patil, N.T.; Kavthe, R.D.; Shinde, V.S. Transition metal-catalyzed addition of C-, N- and O-nucleophiles to unactivated C–C multiple bonds. *Tetrahedron* **2012**, *68*, 8079–8146. [CrossRef]
4. Hintermann, L. Recent developments in metal-catalyzed additions of oxygen nucleophiles to alkenes and alkynes. *Top. Organomet. Chem.* **2010**, *31*, 123–155.
5. Bruneau, C. Group 8 metals-catalyzed O–H bond addition to unsaturated molecules. *Top. Organomet. Chem.* **2013**, *43*, 203–230.
6. Abbiati, G.; Beccalli, E.M.; Rossi, E. Groups 9 and 10 metals-catalyzed O-H bond addition to unsaturated molecules. *Top. Organomet. Chem.* **2013**, *43*, 231–290.
7. Rao, Y.S. Recent advances in the chemistry of unsaturated lactones. *Chem. Rev.* **1976**, *76*, 625–694. [CrossRef]
8. Laduwahetty, T. Saturated and unsaturated lactones. *Contemp. Org. Synth.* **1995**, *2*, 133–149. [CrossRef]
9. Libiszewska, K. Lactones as biologically active compounds. *Biotechnol. Food Sci.* **2011**, *75*, 45–53.
10. Janecki, T. (Ed.) *Natural Lactones and Lactams: Synthesis, Occurrence and Biological Activity*; Wiley-VCH: Weinheim, Germany, 2013; ISBN 9783527334148.
11. Neaţu, F.; Toullec, P.Y.; Michelet, V.; Pârvulescu, V.I. Heterogeneous Au and Rh catalysts for the cycloisomerization reactions of γ-acetylenic carboxylic acids. *Pure Appl. Chem.* **2009**, *81*, 2387–2396. [CrossRef]
12. Bruneau, C.; Neveux, M.; Kabouche, Z.; Ruppin, C.; Dixneuf, P.H. Ruthenium-catalyzed additions to alkynes: Synthesis of activated esters and their use in acylation reactions. *Synlett* **1991**, *11*, 755–763. [CrossRef]
13. Kumar, M.; Bagchi, S.; Sharma, A. The first vinyl acetate mediated organocatalytic transesterification of phenols: A step towards sustainability. *New J. Chem.* **2015**, *39*, 8329–8336. [CrossRef]
14. Liu, X.; Coutelier, O.; Harrison, S.; Tassaing, T.; Marty, J.-D.; Destarac, M. Enhanced solubility of polyvinyl esters in scCO$_2$ by means of vinyl trifluorobutyrate monomer. *ACS Macro Lett.* **2015**, *4*, 89–93. [CrossRef]
15. Foarta, F.; Landis, C.R. Condensation oligomers with sequence control but without coupling reagents and protecting groups via asymmetric hydroformylation and hydroacyloxylation. *J. Org. Chem.* **2016**, *81*, 11250–11255. [CrossRef] [PubMed]
16. Jena, R.K.; Das, U.K.; Ghorai, A.; Bhattacharjee, M. Ruthenium-catalyzed addition of carboxylic acids to progargylic alcohols: An easy route to O-dienyl esters and their tandem atom-transfer radical polymerization. *Eur. J. Org. Chem.* **2016**, 6015–6021. [CrossRef]
17. Konrad, T.M.; Schmitz, P.; Leitner, W.; Franciò, G. Highly enantioselective Rh-catalysed hydrogenation of 1-alkyl vinyl esters using phosphine-phosphoramidite ligands. *Chem. Eur. J.* **2013**, *19*, 13299–13303. [CrossRef] [PubMed]
18. González-Liste, P.J.; León, F.; Arribas, I.; Rubio, M.; García-Garrido, S.E.; Cadierno, V.; Pizzano, A. Highly stereoselective synthesis and hydrogenation of (Z)-1-alkyl-2-arylvinyl acetates: A wide scope procedure for the preparation of chiral homobenzylic esters. *ACS Catal.* **2016**, *6*, 3056–3060. [CrossRef]
19. León, F.; González-Liste, P.J.; García-Garrido, S.E.; Arribas, I.; Rubio, M.; Cadierno, V.; Pizzano, A. Broad scope synthesis of ester precursors of nonfunctionalized chiral alcohols based on the asymmetric hydrogenation of α,β-dialkyl-, α,β-diaryl-, and α-alkyl-β-aryl-vinyl esters. *J. Org. Chem.* **2017**, *82*, 5852–5867. [CrossRef] [PubMed]
20. Takeno, M.; Kikuchi, S.; Morita, S.-I.; Nishiyama, Y.; Ishii, Y. A new coupling reaction of vinyl esters with aldehydes catalyzed by organosamarium compounds. *J. Org. Chem.* **1995**, *60*, 4974–4975. [CrossRef]
21. Goossen, L.J.; Paetzold, J. Decarbonylative Heck olefination of enol esters: Salt-free and environmentally friendly access to vinyl arenes. *Angew. Chem. Int. Ed.* **2004**, *43*, 1095–1098. [CrossRef] [PubMed]
22. Geibel, I.; Dierks, A.; Schmidtmann, M.; Christoffers, J. Formation of δ-lactones by cerium-catalyzed, Baeyer-Villiger-type coupling of β-oxoesters, enol acetates, and dioxygen. *J. Org. Chem.* **2016**, *81*, 7790–7798. [CrossRef] [PubMed]

23. González-Liste, P.J.; Francos, J.; García-Garrido, S.E.; Cadierno, V. The intermolecular hydro-oxycarbonylation of internal alkynes: Current state of the art. *Arkivoc* **2018**, *Part II*, 17–39.

24. Cornils, B.; Herrmann, W.A. (Eds.) *Aqueous-Phase Organometallic Catalysis*, 2nd ed.; Wiley-VCH: Weinheim, Germany, 2004; ISBN 3527307125.

25. Li, C.-J.; Chan, T.-H. *Comprehensive Organic Reactions in Aqueous Media*, 2nd ed.; John Wiley & Sons: Hoboken, NJ, USA, 2007; ISBN 9780471761297.

26. Dixneuf, P.H.; Cadierno, V. (Eds.) *Metal-Catalyzed Organic Reactions in Water*; Wiley-VCH: Weinheim, Germany, 2013; ISBN 9783527331888.

27. Benaglia, M. (Ed.) *Recoverable and Recyclable Catalysts*; John Wiley & Sons: Chichester, UK, 2009; ISBN 9780470681954.

28. Chen, L.; Li, C.-J. Catalytic reactions of alkynes in water. *Adv. Synth. Catal.* **2006**, *348*, 1459–1484. [CrossRef]

29. Wakabayashi, T.; Ishii, Y.; Ishikawa, K.; Hidai, M. A novel catalyst with a cuboidal $PdMo_3S_4$ core for the cyclization of alkynoic acids to enol lactones. *Angew. Chem. Int. Ed.* **1996**, *35*, 2123–2124. [CrossRef]

30. García-Álvarez, J.; Díez, J.; Vidal, C. Pd(II)-catalyzed cycloisomerisation of γ-alkynoic acids and one-pot tandem cycloisomerisation/CuAAC reactions in water. *Green Chem.* **2012**, *14*, 3190–3196. [CrossRef]

31. Phillips, A.D.; Gonsalvi, L.; Romerosa, A.; Vizza, F.; Peruzzini, M. Coordination chemistry of 1,3,5-triaza-7-phosphaadamantane (PTA): Transition metal complexes and related catalytic, medicinal and photoluminescent applications. *Coord. Chem. Rev.* **2004**, *248*, 955–993. [CrossRef]

32. Bravo, J.; Bolaño, S.; Gonsalvi, L.; Peruzzini, M. Coordination chemistry of 1,3,5-triaza-7-phosphaadamantane (PTA) and derivatives. Part II. The quest for tailored ligands, complexes and related applications. *Coord. Chem. Rev.* **2010**, *254*, 555–607. [CrossRef]

33. Ogata, K.; Sasano, D.; Yokoi, T.; Isozaki, K.; Seike, H.; Takaya, H.; Nakamura, M. Pd-complex bound amino acid-based supramolecular gel catalyst for intramolecuar addition-cyclization of alkynoic acids in water. *Chem. Lett.* **2012**, *41*, 498–500. [CrossRef]

34. Hamasaka, G.; Uozumi, Y. Cyclization of alkynoic acids in water in the presence of a vesicular self-assembled amphiphilic pincer palladium complex catalyst. *Green Chem.* **2014**, *50*, 14516–14518. [CrossRef] [PubMed]

35. Michelet, V.; Toullec, P.Y.; Genêt, J.-P. Cycloisomerization of 1,*n*-enynes: Challenging metal-catalyzed rearrangements and mechanistic insights. *Angew. Chem. Int. Ed.* **2008**, *47*, 4268–4315. [CrossRef] [PubMed]

36. Soriano, E.; Marco-Contelles, J. Mechanistic insights on the cycloisomerization of polyunsaturated precursors catalyzed by platinum and gold complexes. *Acc. Chem. Res.* **2009**, *42*, 1026–1036. [CrossRef] [PubMed]

37. Ke, D.; Espinosa, N.A.; Mallet-Ladeira, S.; Monot, J.; Martin-Vaca, B.; Bourissou, D. Efficient synthesis of unsaturated δ- and ε-lactones/lactams by catalytic cycloisomerization: When Pt outperforms Pd. *Adv. Synth. Catal.* **2016**, *358*, 2324–2331, and references therein. [CrossRef]

38. Alemán, J.; del Solar, V.; Navarro-Ranninger, C. Anticancer platinum complexes as non-innocent compounds for catalysis in aqueous media. *Chem. Commun.* **2010**, *46*, 454–456. [CrossRef] [PubMed]

39. Alemán, J.; del Solar, V.; Cubo, L.; Quiroga, A.G.; Navarro-Ranninger, C. New reactions of anticancer-platinum complexes and their intriguing behavior under various experimental conditions. *Dalton Trans.* **2010**, *39*, 10601–10607. [CrossRef] [PubMed]

40. Hashmi, A.S.K.; Toste, F.D. (Eds.) *Modern Gold Catalyzed Synthesis*; Wiley-VCH: Weinheim, Germany, 2012; ISBN 9783527319527.

41. Toste, F.D.; Michelet, V. (Eds.) *Gold Catalysis: An Homogeneous Approach*; Imperial College Press: London, UK, 2014; ISBN 9781848168527.

42. Genin, E.; Toullec, P.Y.; Antoniotti, S.; Brancour, C.; Genêt, J.-P.; Michelet, V. Room temperature Au(I)-catalyzed *exo*-selective cycloisomerization of acetylenic acids: An entry to functionalized γ-lactones. *J. Am. Chem. Soc.* **2006**, *128*, 3112–3113. [CrossRef] [PubMed]

43. Toullec, P.Y.; Genin, E.; Antoniotti, S.; Genêt, J.-P.; Michelet, V. Au_2O_3 as a stable and efficient catalyst for the selective cycloisomerization of γ-acetylenic carboxylic acids to γ-alkylidene-γ-butyrolactones. *Synlett* **2008**, 707–711.

44. Tomás-Mendivil, E.; Toullec, P.Y.; Díez, J.; Conejero, S.; Michelet, V.; Cadierno, V. Cycloisomerization versus hydration reactions in aqueous media: A Au(III)-NHC catalyst that makes the difference. *Org. Lett.* **2012**, *14*, 2520–2523. [CrossRef] [PubMed]

45. Tomás-Mendivil, E.; Toullec, P.Y.; Borge, J.; Conejero, S.; Michelet, V.; Cadierno, V. Water-soluble gold(I) and gold(III) complexes with sulfonated *N*-heterocyclic carbene ligands: Synthesis, characterization, and application in the catalytic cycloisomerization of γ-alkynoic acids into enol-lactones. *ACS Catal.* **2013**, *3*, 3086–3098. [CrossRef]

46. Ferré, M.; Cattoën, X.; Wong Chi Man, M.; Pleixats, R. Sol-gel immobilized *N*-heterocyclic carbene gold complex as a recyclable catalyst for the rearrangement of allylic esters and the cycloisomerization of γ-alkynoic acids. *ChemCatChem* **2016**, *8*, 2824–2831. [CrossRef]

47. Belger, K.; Krause, N. Smaller, faster, better: Modular synthesis of unsymmetrical ammonium salt-tagged NHC-gold(I) complexes and their application as recyclable catalysts in water. *Org. Biomol. Chem.* **2015**, *13*, 8556–8560. [CrossRef] [PubMed]

48. Rodríguez-Álvarez, M.J.; Vidal, C.; Díez, J.; García-Álvarez, J. Introducing deep eutectic solvents as biorenewable media for Au(I)-catalysed cycloisomerisation of γ-alkynoic acids: An unprecedented catalytic system. *Chem. Commun.* **2014**, *50*, 12927–12929. [CrossRef] [PubMed]

49. Mindt, T.L.; Schibli, R. Cu(I)-catalyzed intramolecular cyclization of alkynoic acids in aqueous media: A "click side reaction". *J. Org. Chem.* **2007**, *72*, 10247–10250. [CrossRef] [PubMed]

50. López-Reyes, M.E.; Toscano, R.A.; López-Cortés, J.G.; Alvarez-Toledano, C. Fast and efficient synthesis of Z-enol-γ-lactones through a cycloisomerization reaction of β-hydroxy-γ-alkynoic acids catalyzed by copper(I) under microwave heating in water. *Asian J. Org. Chem.* **2015**, *4*, 545–551. [CrossRef]

51. Li, S.; Jia, W.; Jiao, N. Copper/iron-cocatalyzed highly selective tandem reactions: Efficient approaches to Z-γ-alkylidene lactones. *Adv. Synth. Catal.* **2009**, *351*, 569–575. [CrossRef]

52. Jia, W.; Li, S.; Yu, M.; Chen, W.; Jiao, N. AgNO$_3$ catalyzed cyclization of propargyl-Meldrum´s acids in aqueous solvent: Highly selective synthesis of Z-γ-alkylidene lactones. *Tetrahedron Lett.* **2009**, *50*, 5406–5408. [CrossRef]

53. Ahmar, S.; Fillion, E. Expedient synthesis of complex γ-butyrolactones from 5-(1-arylalkylidene) Meldrum´s acids via sequential conjugate alkynylation/Ag(I)-catalyzed lactonization. *Org. Lett.* **2014**, *16*, 5748–5751. [CrossRef] [PubMed]

54. Liu, W.; Xu, D.D.; Repič, O.; Blacklock, T.J. A mild method for ring-opening aminolysis of lactones. *Tetrahedron Lett.* **2001**, *42*, 2439–2441. [CrossRef]

55. Guo, W.; Gómez, J.E.; Martínez-Rodríguez, L.; Bandeira, N.A.G.; Bo, C.; Kleij, A.W. Metal-free synthesis of *N*-aryl amides using organocatalytic ring-opening aminolysis of lactones. *ChemSusChem* **2017**, *10*, 1969–1975. [CrossRef] [PubMed]

56. Yang, T.; Campbell, L.; Dixon, D.J. A Au(I)-catalyzed *N*-acyl iminium ion cyclization cascade. *J. Am. Chem. Soc.* **2007**, *129*, 12070–12071. [CrossRef] [PubMed]

57. Feng, E.; Zhuo, Y.; Zhang, D.; Zhang, L.; Sun, H.; Jiang, H.; Liu, H. Gold(I)-catalyzed tandem transformation: A simple approach for the synthesis of pyrrolo/pyrido[2,1-*a*][1,3]benzoxazinones and pyrrolo/pyrido[2,1-*a*]quinazolinones. *J. Org. Chem.* **2010**, *75*, 3274–3282. [CrossRef] [PubMed]

58. Patil, N.T.; Lakshmi, P.G.V.V.; Sridhar, B.; Patra, S.; Bhadra, M.P.; Patra, C.R. New linearly and angularly fused quinazolinones: Synthesis through gold(I)-catalyzed cascade reactions and anticancer activities. *Eur. J. Org. Chem.* **2012**, 1790–1799. [CrossRef]

59. Li, Z.; Li, J.; Yang, N.; Chen, Y.; Zhou, Y.; Ji, X.; Zhang, L.; Wang, J.; Xie, X.; Liu, H. Gold(I)-catalyzed cascade approach for the synthesis of tryptamine-based polycyclic privileged scaffolds as α$_1$-adrenergic receptor antagonists. *J. Org. Chem.* **2013**, *76*, 10802–10811. [CrossRef] [PubMed]

60. Feng, E.; Zhou, Y.; Zhao, F.; Chen, X.; Zheng, L.; Jiang, H.; Liu, H. Gold-catalyzed tandem reaction in water: An efficient and convenient synthesis of fused polycyclic indoles. *Green Chem.* **2012**, *14*, 1888–1895. [CrossRef]

61. Zhuo, Y.; Zhai, Y.; Ji, X.; Liu, G.; Feng, E.; Ye, D.; Zhao, L.; Jiang, H.; Liu, H. Gold(I)-catalyzed one-pot tandem coupling/cyclization: An efficient synthesis of pyrrolo-/pyrido[2,1-*b*]benzo[*d*][1,3]oxazin-1-ones. *Adv. Synth. Catal.* **2010**, *352*, 373–378. [CrossRef]

62. Zhou, L.; Jiang, H.-F. Synthesis of phthalides via Pd/CNTs-catalyzed reaction of terminal alkynes and *o*-iodobenzoic acid under copper- and ligand-free conditions. *Tetrahedron Lett.* **2007**, *48*, 8449–8452. [CrossRef]

63. García-Álvarez, J.; Díez, J.; Gimeno, J. A highly efficient copper(I) catalyst for the 1,3-dipolar addition of azides with terminal and 1-iodoalkynes in water: Regioselective synthesis of 1,4-disubstituted and 1,4,5-trisubstituted 1,2,3-triazoles. *Green Chem.* **2010**, *12*, 2127–2130. [CrossRef]

64. Denard, C.A.; Hartwig, J.F.; Zhao, H. Multistep one-pot reactions combining biocatalysts and chemical catalysts for asymmetric synthesis. *ACS Catal.* **2013**, *3*, 2856–2864. [CrossRef]
65. Gröger, H.; Hummel, W. Combining the 'two worlds' of chemocatalysis and biocatalysis towards multi-step one-pot processes in aqueous media. *Curr. Opin. Chem. Biol.* **2014**, *19*, 171–179. [CrossRef] [PubMed]
66. Wallace, S.; Balskus, E.P. Opportunities for merging chemical and biological synthesis. *Curr. Opin. Biotechnol.* **2014**, *30*, 1–8. [CrossRef] [PubMed]
67. Rodríguez-Álvarez, M.J.; Ríos-Lombardía, N.; Schumacher, S.; Pérez-Iglesias, D.; Morís, F.; Cadierno, V.; García-Álvarez, J.; González-Sabín, J. Combination of metal-catalyzed cycloisomerizations and biocatalysis in aqueous media: Asymmetric construction of chiral alcohols, lactones and γ-hydroxy-carbonyl compounds. *ACS Catal.* **2017**, *7*, 7753–7759. [CrossRef]
68. Cadierno, V.; Francos, J.; Gimeno, J. Ruthenium(IV)-catalyzed Markovnikov addition of carboxylic acids to terminal alkynes in aqueous medium. *Organometallics* **2011**, *30*, 852–862. [CrossRef]
69. Chanda, A.; Fokin, V.V. Organic synthesis "on water". *Chem. Rev.* **2009**, *109*, 725–748. [CrossRef] [PubMed]
70. Butler, R.N.; Coyne, A.G. Water: Nature's reaction enforcer—Comparative effects for organic synthesis "in-water" and "on-water". *Chem. Rev.* **2010**, *110*, 6302–6337. [CrossRef] [PubMed]
71. Nicks, F.; Aznar, R.; Sainz, D.; Muller, G.; Demonceau, A. Novel, highly efficient and selective ruthenium catalysts for the synthesis of vinyl esters from carboxylic acids and alkynes. *Eur. J. Org. Chem.* **2009**, 5020–5027. [CrossRef]
72. Kleman, P.; González-Liste, P.J.; García-Garrido, S.E.; Cadierno, V.; Pizzano, A. Highly enantioselective hydrogenation of 1-alkylvinyl benzoates: A simple, nonenzymatic access to chiral 2-alkanols. *Chem. Eur. J.* **2013**, *19*, 16209–16212. [CrossRef] [PubMed]
73. Kleman, P.; González-Liste, P.J.; García-Garrido, S.E.; Cadierno, V.; Pizzano, A. Asymmetric hydrogenation of 1-alkyl and 1-aryl vinyl benzoates: A broad scope procedure for the highly enantioselective synthesis of 1-substituted ethyl benzoates. *ACS Catal.* **2014**, *4*, 4398–4408. [CrossRef]
74. González-Liste, P.J.; García-Garrido, S.E.; Cadierno, V. Gold(I)-catalyzed addition of carboxylic acids to internal alkynes in aqueous medium. *Org. Biomol. Chem.* **2017**, *15*, 1670–1679. [CrossRef] [PubMed]
75. Scheid, G.; Kuit, W.; Ruijter, E.; Orru, R.V.A.; Henke, E.; Bornscheuer, U.; Wessjohann, L.A. A new route to protected acyloins and their enzymatic resolution with lipases. *Eur. J. Org. Chem.* **2004**, 1063–1074. [CrossRef]
76. Carpino, P.A.; Griffith, D.A.; Sakya, S.; Dow, R.L.; Black, S.C.; Hadcock, J.R.; Iredale, P.A.; Scott, D.O.; Fichtner, M.W.; Rose, C.R.; et al. New bicyclic cannabinoid receptor-1 (CB$_1$-R) antagonists. *Bioorg. Med. Chem. Lett.* **2006**, *16*, 731–736. [CrossRef] [PubMed]
77. Lu, H.-L.; Wu, Z.-W.; Song, S.-Y.; Liao, X.-D.; Zhu, Y.; Huang, Y.-S. An improved synthesis of nomegestrol acetate. *Org. Process Res. Dev.* **2014**, *18*, 431–436. [CrossRef]
78. Kaila, N.; Janz, K.; DeBernardo, S.; Bedard, P.W.; Camphausen, R.T.; Tam, S.; Tsao, D.H.H.; Keith, J.C.; Nickerson-Nutter, C.; Shilling, A.; Young-Sciame, R.; Wang, Q. Synthesis and biological evaluation of quinoline salicylic acids as P-Selectin antagonists. *J. Med. Chem.* **2007**, *50*, 21–39. [CrossRef] [PubMed]
79. Ruiz, J.F.M.; Radics, G.; Windle, H.; Serra, H.O.; Simplício, A.L.; Kedziora, K.; Fallon, P.G.; Kelleher, D.P.; Gilmer, J.F. Design, synthesis, and pharmacological effects of a cyclization-activated steroid prodrug for colon targeting in inflammatory bowel disease. *J. Med. Chem.* **2009**, *52*, 3205–3211. [CrossRef] [PubMed]
80. Jeschke, J.; Korb, M.; Rüffer, T.; Gäbler, C.; Lang, H. Atom economic ruthenium-catalyzed synthesis of bulky β-oxo esters. *Adv. Synth. Catal.* **2015**, *357*, 4069–4081, and references therein. [CrossRef]
81. Mitsudo, T.; Hori, Y.; Yamakawa, Y.; Watanabe, Y. Ruthenium-catalyzed selective addition of carboxylic acids to alkynes. A new synthesis of enol esters. *J. Org. Chem.* **1987**, *52*, 2230–2239. [CrossRef]
82. Bauer, E. Transition-metal-catalyzed functionalization of propargylic alcohols and their derivatives. *Synthesis* **2012**, *44*, 1131–1151. [CrossRef]
83. Cadierno, V.; Francos, J.; Gimeno, J. Ruthenium-catalyzed synthesis of β-oxo esters in aqueous medium: Scope and limitations. *Green Chem.* **2010**, *12*, 135–143. [CrossRef]

 *catalysts*

Review

Ultrasonic Monitoring of Biocatalysis in Solutions and Complex Dispersions

Vitaly Buckin * and Margarida Caras Altas

School of Chemistry, University College Dublin, Belfield, Dublin 4, Ireland; maguy.ca@gmail.com
* Correspondence: vitaly.buckin@ucd.ie; Tel.: +353-1-7162371

Received: 19 September 2017; Accepted: 26 October 2017; Published: 15 November 2017

Abstract: The rapidly growing field of chemical catalysis is dependent on analytical methods for non-destructive real-time monitoring of chemical reactions in complex systems such as emulsions, suspensions and gels, where most analytical techniques are limited in their applicability, especially if the media is opaque, or if the reactants/products do not possess optical activity. High-resolution ultrasonic spectroscopy is one of the novel technologies based on measurements of parameters of ultrasonic waves propagating through analyzed samples, which can be utilized for real-time non-invasive monitoring of chemical reactions. It does not require optical transparency, optical markers and is applicable for monitoring of reactions in continuous media and in micro/nano bioreactors (e.g., nanodroplets of microemulsions). The technology enables measurements of concentrations of substrates and products over the whole course of reaction, analysis of time profiles of the degree of polymerization and molar mass of polymers and oligomers, evolutions of reaction rates, evaluation of kinetic mechanisms, measurements of kinetic and equilibrium constants and reaction Gibbs energy. It also provides tools for assessments of various aspects of performance of catalysts/enzymes including inhibition effects, reversible and irreversible thermal deactivation. In addition, ultrasonic scattering effects in dispersions allow real-time monitoring of structural changes in the medium accompanying chemical reactions.

Keywords: ultrasonic spectroscopy; HR-US; ultrasonic velocity; ultrasonic attenuation; biocatalysis; enzymes; metal surface catalysis; degree of polymerization; average molar mass; microemulsions

---

1. Introduction

The availability of efficient methods for real-time non-destructive monitoring of catalytic transformation between reactants and products in different reaction media, from solutions to emulsion, suspensions, or gels, is an important factor in the modern development of catalysis and its applications in industrial processes [1,2]. A variety of 'electromagnetic' spectroscopic techniques such as infrared, Raman, fluorescence, and UV-Vis, can provide real-time data on concentrations of reactant and product under varying reaction conditions [1,3–7]. However, the efficiency of these techniques is dependent on the optical transparency of medium and can be affected by light scattering in dispersions. In general, these techniques have a limited dynamic (concentration) range. In addition, UV-Vis and fluorescence spectroscopies require an optical activity of the reactants or products and, in their absence, the analysis is more complicated as chromogenic and fluorogenic substrates are needed [8–10]. Therefore, discontinuous methods are widely used for the monitoring of chemical reactions, including chromatography, mass spectrometry (MS) and capillary electrophoresis [9]. These methods involve the collection of samples from a reaction mixture, and often include invasive procedures of stepwise extractions or sample pretreatments [11], which is time-consuming and costly.

This paper describes applications of an alternative to 'electromagnetic' spectroscopy-ultrasonic spectroscopy for precision real-time non-invasive monitoring of chemical reactions in solutions and complex dispersions. Ultrasonic spectroscopy employs high-frequency (MHz range) waves of compressions and decompressions (longitudinal deformations), which probe the elastic properties of materials determined by the intermolecular interactions/forces and the microstructural organization [12–14]. The two major parameters measured in ultrasonic spectroscopy are ultrasonic attenuation, α, and ultrasonic velocity, u. Attenuation represents the exponential decay of the amplitude of the ultrasonic wave, A, of the oscillations of pressure (or longitudinal deformation) with distance travelled, z:

$$A = A_0 e^{-\alpha z} \cos[2\pi f(\frac{z}{u} - t)] \tag{1}$$

where t is time and f is the frequency of the wave. The attenuation, α, is determined by the energy losses in compressions and decompressions in ultrasonic waves. Monitoring of ultrasonic attenuation provides information on fast (relaxation time is the order of $(2\pi f)^{-1}$) dynamics of molecular processes [15] and microstructural (down to nm scale) organization of materials, including particle sizing in emulsions and suspensions (nm and μm scales), as well as characterization of aggregation, gelation, crystallization and creaming [12,16–19].

In liquids with a limited level of ultrasonic attenuation, the ultrasonic velocity is determined by the adiabatic compressibility $\beta_S = -\frac{1}{V}\left(\frac{\partial V}{\partial P}\right)_S$, where V is the volume, P is the pressure and S the entropy, and by the density, ρ, of the medium:

$$u = \frac{1}{\sqrt{\beta_S \rho}} \tag{2}$$

Compressibility, β_S, is extremely sensitive to the molecular organization and intermolecular interactions in the medium. This can be applied in analysis of a broad range of molecular processes, including catalyzed chemical reactions, if the measurements of ultrasonic velocity are performed with sufficient accuracy [2,13,20,21]. Although ultrasonic spectroscopy has been utilized for material analysis for a long time and has demonstrated various successful applications [12,14,22], the capability of this technique in analysis of chemical reactions has been restricted by a number of factors. These include limited resolution/precision in measurements of ultrasonic parameters, the requirement for large sample volumes and often complicated measuring procedures. The development of high-resolution ultrasonic spectroscopy (HR-US), based on advances in the principles of ultrasonic measurements, electronics and digital signal processing, surpass these limitations [23]. HR-US instruments (Figure 1) enable ultrasonic measurements with exceptionally high precision, down to 0.2 mm/s for ultrasonic velocity, in a broad range of sample volumes, 0.03 mL (droplet size) to several mL with a typical frequency range 1 to 20 MHz [23,24]. High precision of this technique allows for the monitoring of small changes in the concentrations of substrates and products in chemical reactions [5,8,25,26]. In addition, the geometry of HR-US ultrasonic cells is optimized for easy filling, refilling, cleaning and sterilization. They can accommodate aggressive liquids such as strong acids or volatile organic solvents, without evaporation throughout the course of measurements. HR-US measurements can be performed in a wide range of temperatures (-40 to 130 °C), at ambient or elevated pressures, in static and flow-through regimes, and in different media ranging from dilute solutions to semisolid materials [23]. The measurements can include automated precision titrations of the analyzed liquids with a titrant [27], and also programmable temperature ramps for temperature profiling [2,28]. These qualities enable the application of HR-US technique for non-destructive real-time monitoring of chemical reactions in a broad range of media and environmental conditions.

The paper reviews the underlying principles and outcomes of application of HR-US technique for real-time monitoring of chemical reactions in continuous media and in nano-bioreactors (nanodroplets) of complex dispersions such as milks, suspensions of protein nanoparticles and microemulsions. This includes precision real-time measurements of concentrations of reactants and products (reaction progress curves), time profiles of the average degree of polymerization and of molar mass of polymers and oligomers, time profiles of the reaction rates, the monitoring of structural rearrangements (particle size) in a course of reactions, evaluation of kinetic mechanisms, measurements of kinetic and equilibrium constants of reactions and reaction Gibbs energy. It also discusses the tools provided by this ultrasonic technology for assessment of various aspects of the performance of catalysts/enzymes including inhibition effects and reversible and irreversible thermal deactivation. Although, most of the examples discussed are the reactions catalyzed by enzymes, the detection principles and the methodologies described shall be applicable for a variety of catalysts in aqueous and non-aqueous liquids.

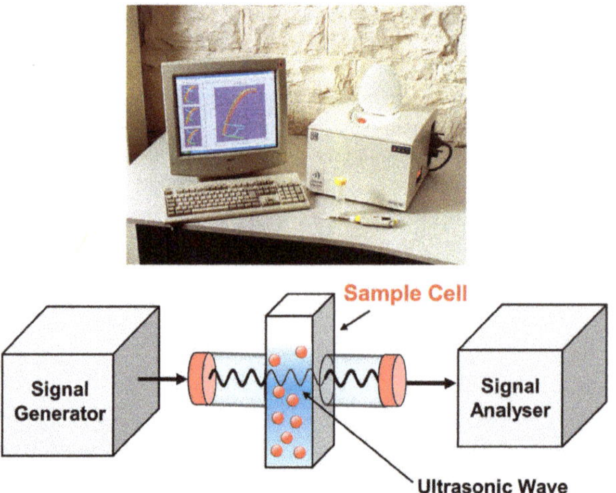

Figure 1. Illustration of operation of high-resolution ultrasonic spectrometers.

2. Monitoring of Chemical Reactions with Ultrasonic Velocity

2.1. Detection Principles. Reactions Progress Curves

Chemical reactions in a liquid mixture are accompanied by a change in the intrinsic properties of molecules involved in the reaction and by a change in their interactions with the environment, which includes solvation effects (hydration in aqueous solutions). This changes the compressibility and density of the mixture, and affects the ultrasonic velocity in the mixture. Thus, monitoring of the evolution of ultrasonic velocity, $u(t)$, with reaction time, t, shall provide real-time information on the concentration of reactants transferred to products. For a chemical reaction:

$$a_1 A_1 + a_2 A_2 + \ldots + a_k A_k \rightarrow b_1 B_1 + b_2 B_2 + \ldots + b_l B_l \tag{R1}$$

involving a_1, a_2, $\ldots$, a_k moles of reactants A_1, A_2, $\ldots$, A_k and producing b_1, b_2, $\ldots$, b_k moles of products B_1, B_2, $\ldots$, B_k the change of ultrasonic velocity within time interval δt, $\delta u(t)$, caused by the reaction can be presented as:

$$\delta u(t) = -u_0 \Delta a_r \delta c(t) \tag{3}$$

where $\delta c(t)$ is the change of concentration of one of the reactants during time interval δt and Δa_r is the concentration increment of ultrasonic velocity of the reaction defined as:

$$\Delta a_r = -\frac{1}{u_0} \frac{du}{dc} \tag{4}$$

here u_0 is the ultrasonic velocity in the reaction medium without reactants and products, and the derivative is taken at the conditions determined by Reaction (R1). The negative sign in the above equation is introduced to compensate for the negative sign of δc over the reaction. In non-concentrated mixtures, in an absence of specific interactions between the reactants and products, the value of Δa_r is expected to be constant during the reaction, especially if the physico-chemical properties of the products are similar to the properties of the reactants. In this case the concentration $c(t)$ and the reaction extent $\zeta(t)$ can be obtained as:

$$c(t) = c^0 - \frac{u(t) - u^0}{u_0 \Delta a_r}$$
$$\zeta(t) = \frac{u(t) - u^0}{u_0 \Delta a_r c^0} \tag{5}$$

where u^0 and c^0 are ultrasonic velocity and the concentration of the reactant in the mixture at the reaction time zero (Reaction (R1) fully shifted to the left) and $u(t)$ is ultrasonic velocity in the mixture at the reaction time t. As all concentrations of reactants and products are linked with each other by the stochiometric relationships of Reaction (R1), the concentration of any of the reactants or products or other parameter defining them (e.g., reaction extent) can be utilized in the above relationships. Although the choice of parameter representing the change of concentration of reactants and products (reaction progress) affects the concentration increment of ultrasonic velocity of the reaction, Δa_r, the value of Δa_r determined for a particular reactant or product can be recalculated into the value of Δa_r for any other reactant or product using the stochiometric relationships of Reaction (R1).

The parameter Δa_r can be presented as the difference of concentration increments of ultrasonic velocity of the reactants, a_R, and products, a_P:

$$\Delta a_r = a_P - a_R$$
$$a_P = \frac{u_P - u_0}{u_0 c^0}; \; a_R = \frac{u_R - u_0}{u_0 c^0} \tag{6}$$

here u_R and u_P are the ultrasonic velocities in the reaction mixture where the equilibrium in the analyzed reaction is shifted fully to the left ($\zeta = 0$) and to the right ($\zeta = 1$), respectively. The presence of the parameter u_0 in Equations (3)–(6) provides direct relationships linking a_R and a_P with apparent compressibility and volume of the reactants and products (see for example Reference [13]). 'Libraries' of these properties can be used for estimations of a_R, a_P and Δa_r (see Discussion in [25] as example). Although the concentration $c(t)$ can be expressed in different units, such as molarity or molality, it is useful to use moles per kg of mixture. In this case, $c(t)$ can be easily obtained from the weight fraction of the relevant component of the reaction, w, often utilized in preparations of mixtures: $c = \frac{w}{M}$, where M is the molar mass in kg/mol. If required, the molarity of a component A_i (or B_i) can be calculated from the concentration c_i (mol per kg of mixture) as:

$$[A_i] = \rho c_i \tag{7}$$

where ρ is the density of the mixture in kg/m^3. The reaction rate, r, can be obtained as a slope of the time profile of c or ζ:

$$r = -\frac{dc}{dt} = -c^0 \frac{d\zeta}{dt} \tag{8}$$

The reaction rate expressed in molarity units at time t can be obtained through multiplication of r by the density of the mixture at that time (t).

For some reactions the value of Δa_r can change in the course of the reaction due to a strong dependence of concentration increments of ultrasonic velocity of reactants and products on their concentration, or other reasons (example: decomposition of H_2O_2). For a broad range of solutions the concentration increments of ultrasonic velocity of solutes show a linear dependence on their concentration, $c(t)$. In this case, the parameter Δa_r can be presented as: $\Delta a_r = \Delta a_0 + \Delta a_1 c(t)$, where

Δa_0 and Δa_1 are constants, which do not depend on $c(t)$. In this case integration of Equation (3) with subsequent solution of the resulted quadratic equation, and selection of an appropriate solution ($c(t = 0) = c^0$) provides the following relationship for calculation of $c(t)$ from ultrasonic velocity profile:

$$c(t) = \frac{\Delta a_0}{\Delta a_1}\left(\sqrt{\left(1 + \frac{\Delta a_1}{\Delta a_0}c^0\right)^2 - 2\frac{\Delta a_1}{\Delta a_0}\frac{u(t) - u^0}{\Delta a_0 u_0}} - 1\right) \tag{9}$$

At $\Delta a_1 \rightarrow 0$ this relationship is reduced to Equation (5).

2.2. Reactions Involving the Same Type of Covalent Bonds in Multiple Reactants and Products

For reactions in mixtures with multiple reactants and products involving formation or breaking of the same type of covalent bonds it is often beneficial to represent the reaction progress curves through the dependence of concentration of the bonds in the mixture as a function of time. Examples could be reactions of the hydrolysis of oligosaccharides (glycosidic bonds) and proteins (peptide bonds) discussed in subsequent chapters. The hydrolysis of these bonds can be presented as:

$$R_1 - O - R_2 + H_2O \rightarrow R_1 - OH + HO - R_2 \tag{R2}$$

and the reaction progress can be described through the evolution of concentration of the $-O-$ bonds in the mixture. In this case, the parameter $c(t)$ in Equations (3)–(8) represents the concentration of $-O-$ bonds, hydrolysable in Reaction (R2). Consequently, the concentration increment of ultrasonic velocity of reaction, Δa_r, represents the relative change of ultrasonic velocity caused by hydrolysis of one mole of $-O-$ bonds in one kg of the mixture. Following Equation (5) the concentration of bonds hydrolyzed, $c_{bh}(t) = c^0 - c(t)$, can be calculated as:

$$c_{bh}(t) = \frac{u(t) - u^0}{u_0 \Delta a_r} \tag{10}$$

The parameter Δa_r is mainly determined by the difference in the hydration (solvation) level and in the intrinsic properties of the atomic groups of the reactants and of the products affected by the reaction [25]. Therefore, for sizable oligomers and for polymers the contribution of the atomic groups of reactants and products positioned outside of the local environment of $-O-$ bond to Δa_r is expected to be very small. Consequently, the value of Δa_r for hydrolysis of these molecules should not depend on the extent of hydrolysis. An exception to this are polymers with well-defined compact structure and high level of cooperativity of structural rearrangements caused by hydrolysis, affecting the hydration characteristics of their atomic groups and the intrinsic compressibility of these molecules.

The concentration of bonds hydrolyzed can be expressed in molarity units (moles of bonds hydrolyzed in one L of mixture, $[bh]$), if the density of the mixture, ρ, is known: $[bh] = \rho c_{bh}(t)$. The ultrasonically measured $[bh]$ could be employed in estimations of the change of osmolarity of solutions caused by the reaction, which is required in various applications. Another usage of the parameter $c_{bh}(t)$ is calculations of the average degree of polymerization and molar mass, discussed in the following chapter.

2.3. Degree of Polymerization and Molar Mass of Linear Polymers or Oligomers

In the case of hydrolysis of a linear polymer or oligomer, the measured concentration of bonds hydrolyzed at any time t of reaction, $c_{bh}(t)$, can be recalculated into the average degree of polymerization at this time, $\overline{DP}$, as well as into the average molar mass, $\overline{M}$ (see Appendix A.1 for definitions). The relationships for calculations of the average degree of polymerization and of molar mass are discussed in Appendix A.1 and are the following:

$$\overline{DP} = \frac{\overline{DP}^0}{1 + c_{bh}(t)\frac{\overline{DP}^0 - 1}{c_b^0}} ; \overline{M} = \frac{\overline{M}^0 - M_{H_2O}}{1 + c_{bh}(t)\frac{\overline{M}^0}{w_p^0}} + M_{H_2O} \tag{11}$$

where $\overline{DP}^0$ and $\overline{M}^0$ are consequently the average degree of polymerization and the average molar mass at time zero ($t = 0$) and w_p^0 is the weight fraction of the polymer or oligomer (kg in one kg of the mixture) at time zero, M_{H_2O} is the molar mass of water. For the concentration $c_{bh}(t)$ in Equation (11) taken in moles per kg (of mixture), the molar mass shall be expressed in kDa (kg per mole). The ratios $\frac{\overline{DP}^0 - 1}{c_b^0}$ and $\frac{\overline{M}^0}{w_p^0}$ are related to each other as: $\frac{\overline{DP}^0 - 1}{c_b^0} = \frac{\overline{M}^0}{w_p^0}$.

2.4. Calibration for Ultrasonic Velocity

Calibration of HR-US technique requires determination of the concentration increment of ultrasonic velocity of reaction, Δa_r, which allows calculations of the reaction progress curves (evolution of concentrations of reactants and products with time) from the measured time profiles of ultrasonic velocity using Equation (5), or (9), or (10). Four major methods of determination of Δa_r are outlined below.

2.4.1. Method 1

The concentration increment of ultrasonic velocity of reaction, Δa_r, can be obtained by measuring the ultrasonic velocity in the mixture containing reactants only, u_R, and velocity in the mixture containing products only, u_P, from which a_R and a_P can be obtained, and Δa_r calculated according to Equation (6). Examples of such measurements are illustrated in Figure 2A–C for hydrolysis of lactose in water and in infant milk, and for decomposition of hydrogen peroxide in water. Such measurements can be utilized for the evaluation of the dependence of Δa_r on the concentration of reactants (or products), which, however, is often small and can be neglected especially in dilute solutions.

In hydrolysis in aqueous solutions water acts as one of the reactants. Therefore, measurements of concentration increment of ultrasonic velocity of reactants, a_R, shall include water added to the reactant in an appropriate equimolar (consumed by the reaction) amount.

In this case, the contribution of reactants to ultrasonic velocity in the mixture, u_R, can be presented as a sum of the addition of other than water reactants ($R1$) to the solvent Δu_{R1} plus the effect of subsequent addition of water, Δu_w: $u_R - u_0 = \Delta u_{R1} + \Delta u_w$. Consequently, $a_R = a_{R1} + a_w$ and $\Delta a_r = (a_P - a_{R1}) - a_w$, where $a_w = \frac{\Delta u_w}{u_0 c^0}$. The parameter Δu_w represents the effect of dilution of the mixture with water. For non-concentrated aqueous solutions, a_w is normally small and can be neglected.

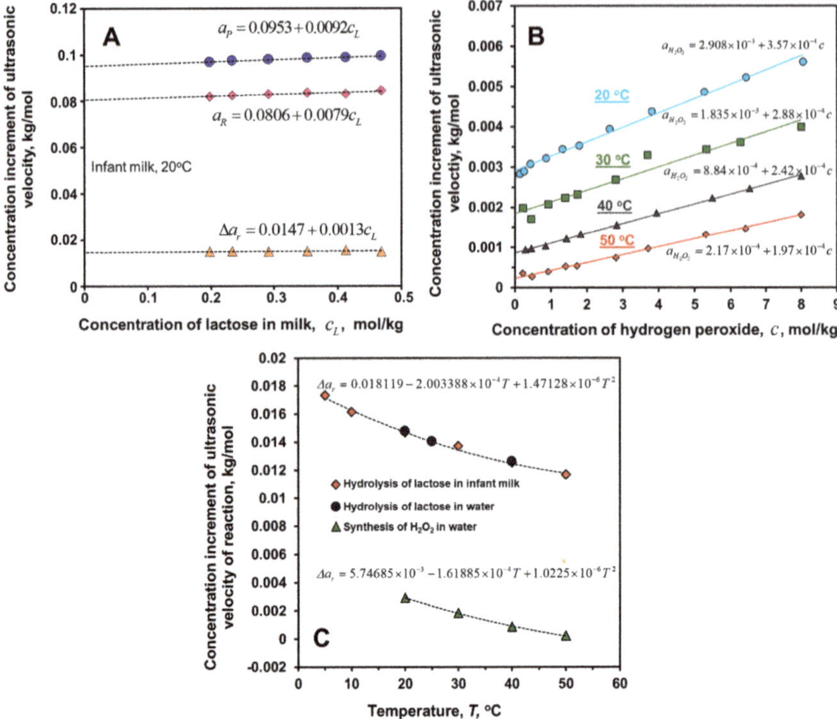

Figure 2. Application of Method 1 for measurements of concentration increments of ultrasonic velocity of reaction, Δa_r.

Figure 2A. *Hydrolysis of lactose in infant milk.*

Reaction (R4). Concentration increments of ultrasonic velocity of reactants, a_R, diamonds, (lactose, 0.198 mol/kg, plus equimolar amount of water), and products, a_P, circles, (glucose, 0.198 mol/kg, plus equimolar amount of galactose), in Cow & Gate First Infant Milk with added lactose at 20 °C at 5 MHz (no significant frequency dependence detected within 2 to 15 MHz). c_L denotes the concentration of lactose in milk prior to the addition of reactants or products. The lowest c_L (0.198 mol/kg) represents milk without added lactose. Concentration increment of ultrasonic velocity of hydrolysis, Δa_r was calculated using Equation (6). The linear dependence of Δa_r on c_L (triangles) was extrapolated to c_L = 0 to account for minor effects of lactose present in milk prior addition of reactants and products producing $\Delta a_r = (0.0147 \pm 0.0004)$ kg/mol [8]. Adapted with permission from Reference [8] Copyright © 2016 American Chemical Society.

Figure 2B. *Decomposition of hydrogen peroxide in water.*

Reaction (R7). The concentration increment of ultrasonic velocity of hydrogen peroxide, $a_{H_2O_2}$, in aqueous solution of different concentration, c, of H_2O_2 at 20 °C, 30 °C, 40 °C and 50 °C at 12 MHz (no significant frequency dependence detected within 2 to 12 MHz) calculated according to Equation (6). As oxygen gas produced in the decomposition of H_2O_2 is removed from the solution and the contribution of the second product, water, is negligibly small (see text) $\Delta a_r \cong -a_{H_2O_2}$. Adapted from Reference [5] with permission from The Royal Society of Chemistry. The linear fit parameters were obtained by fitting of the dependence of $a_{H_2O_2}$ on concentration of H_2O_2, which provides slightly better extrapolation of $a_{H_2O_2}$ to infinite dilution than the direct fitting of ultrasonic velocity vs. concentration utilized in [5].

Figure 2C. *Examples of temperature dependence of* Δa_r.

Red diamonds: hydrolysis of lactose in infant milk and in water [8]. Blue circles: hydrolysis of lactose in water [8]. Green triangles: decomposition of hydrogen peroxide in water at infinite dilution (calculated from plots (B)). For illustrative purposes, the negative value of concentration increment of ultrasonic velocity of decomposition of hydrogen peroxide, $-\Delta a_r$, is plotted (see Discussion in Section 6.5).

2.4.2. Method 1 in Hydrolysis of Oligomers and Polymers

For Reaction (R2) of hydrolysis of the same type of covalent bond (and same Δa_r) in a polymer and an oligomer, the concentration increment of ultrasonic velocity of the reaction (expressed through relative change of ultrasonic velocity in hydrolysis of one mole of bonds hydrolyzed in one kg of mixture) can be obtained through the measurements of ultrasonic velocity u_P in the mixture containing the products $R_1 - OH$ and $HO - R_2$ only (Reaction (R2) is fully shifted to the right) and the velocity, u_R, in the mixture containing the reactants only, (polymer or oligomer) $R_1 - O - R_2$ plus equimolar amount of water (Reaction (R2) is fully shifted to the left). The concentration c^0 in Equation (6) shall represent the concentration of the $-O-$ bonds at time zero.

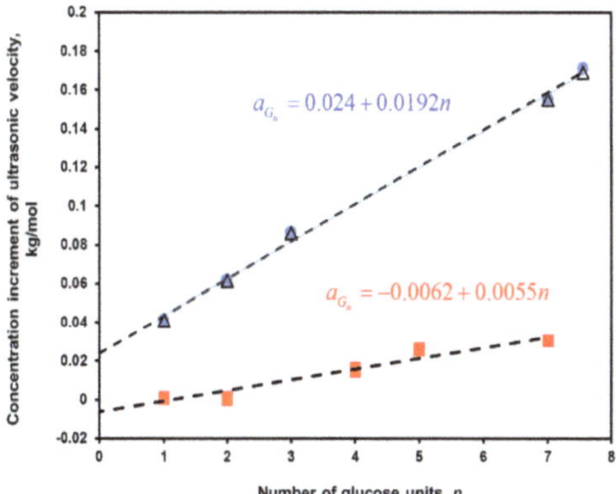

Figure 3. Concentration increments of ultrasonic velocity of maltodextrins, a_{G_n}, in aqueous solution and in isopropilmiristate microemulsion at 25 °C.
Blue circles: a_{G_n} of glucose oligomers in water at 25 °C at concentration 0.02 kg/kg calculated according to Equation (6) from ultrasonic velocities at 8 MHz (no significant frequency dependence detected within the analyzed frequency range 5 to 8 MHz).
Triangles: Same as above at concentration 0.01 kg/kg.
Red squares: a_{G_n} of glucose oligomers in the aqueous droplets of IPM microemulsion system (water (24% kg/kg), oil (isopropilmiristate (IPM), 38% kg/kg), cosurfactant (*n*-propanol, 19% kg/kg), surfactant (E200, 19% kg/kg)) at concentration 0.025 kg per kg of the aqueous phase (overall concentration 0.06 kg/kg), calculated according to Equation (6) from ultrasonic velocities at 8 MHz (no significant frequency dependence detected within the analyzed frequency range 3 to 8 MHz). Details of preparation of the microemulsion were described earlier [29–31]. $n = 1$—glucose, $n = 2$—maltose, $n = 3$—maltotriose, $n = 4$—maltotetrose, $n = 5$—maltopentose, $n = 7$—maltoheptaose, and average $n = 7.55$—maltodextrin (provided by manufacturer and confirmed with bicinchoninic acid (BCA) assay [32,33]).

For oligomers and polymers consisting of the same monomeric units another approach can be utilized if the concentration increments of ultrasonic velocity for homologous series are available. An example of such data for maltodextrin (oligomers of glucose) is illustrated in Figure 3. According to the figure the concentration increment of an oligomer G_n, composed of n molecules of glucose is represented as: $a_{G_n} = b_0 + b_1 n$, where b_0 and b_1 are constants. Accordingly, the change of concentration increment of ultrasonic velocity per one bond hydrolyzed in the reaction $G_{k+l} + H_2O \rightarrow G_k + G_l$ is given as $\Delta a_r = (a_{G_k} + a_{G_l}) - (a_{G_{k+l}} + a_w)$, which produces:

$$\Delta a_r = b_0 - a_w \tag{12}$$

The parameter b_0 in this equation represents the intercept of the line drawn trough the points a_{G_n} vs. n with the Y axis at $n = 0$ (Figure 3). The results of the application of Method 1 for hydrolysis of maltodextrin in aqueous solution and in aqueous droplets of microemulsion are outlined in Table 1.

Table 1. Examples of concentration increments of ultrasonic velocity of reaction, Δa_r, kg/mol.

Reaction	Method 1	Method 2	Method 3	Conditions
Hydrolysis of lactose, Reactions (R2, R4)	$\cong 0.018119$ $-2.003388 \times 10^{-4}\,T$ $+1.47128 \times 10^{-6}\,T^2$	-	-	Infant milk, temperature, T, 10 to 50 °C [a]
Hydrolysis of lactose, Reactions (R2, R4)	0.0144 ± 0.0002	0.0136 ± 0.0006	-	Dilute aqueous solutions (0 to 0.2 kg/kg), 20 °C [a]
Synthesis of ATP, Reaction (R6)	-	-0.0306 ± 0.0009	-0.0310 ± 0.0003	Dilute aqueous solutions, of ADP, 15 mM of phosphocreatine in 50 mM gly-gly buffer, 0.02% BSA, 5 mM Mg acetate, 25 °C, pH 7.4
Hydrolysis of maltodextrin, Reactions (R2, R3)	0.0240 ± 0.0002	-	-	Dilute aqueous solutions (0.01 and 0.02 kg/kg), 25 °C
Hydrolysis of maltodextrin, Reactions (R2, R3)	-0.0093 ± 0.0008	-	-	IPM microemulsion, 25 °C, 0.025 kg/kg in aqueous phase
Hydrolysis of protein, β-lactoglobulin, by α-chymotrypsin, Reactions (R2, R5)	-	0.0700 ± 0.0015	-	Dilute aqueous solutions (0 to 0.01 kg/kg), 25 °C, 0.1 mol/L potassium phosphate buffer, pH 7.8
Decomposition of hydrogen peroxide, Reaction (R7)	$\cong -5.74685 \times 10^{-3}$ $+1.61885 \times 10^{-4}\,T$ $-1.0225 \times 10^{-6}\,T^2$	-	-	Infinite dilution in water (for concentrated solutions see Figure 2B), temperature, T, 20 to 50 °C
Hydrolysis of cellobiose, Reactions (R2, R8)	0.0126 ± 0.0002	0.013 ± 0.001	-	Dilute aqueous solutions (0 to 0.1 kg/kg), 50 °C [b]

T is temperature in Celsius; [a] Reference [8]; [b] Reference [25].

2.4.3. Method 2

An alternative way to determine Δa_r involves simultaneous measurements of the change of ultrasonic velocity during the analyzed reaction and the measurements of the concentration $c(t)$, using an appropriate discontinuous technique at different reaction times. In this case according to Equation (5) the slop of the line representing the plot of $u(t) - u^0$ vs. $c_P(t) = c^0 - c(t)$, S, provides:

$$\Delta a_r = \frac{S}{u_0} \tag{13}$$

Examples of the application of discontinuous techniques for hydrolysis of peptide bonds of β-lactoglobulin, hydrolysis of lactose and synthesis of ATP are given in Figure 4A. The figure shows an increase of ultrasonic velocity with the reaction extent for the first two reactions. The primary reason for this is the higher level of hydration of atomic groups of the products when compared with the reactants [8,25,35]. The hydration effects reduce the compressibility of the water surrounding the atomic groups of reactants and products during reactions (see for example [13,25,35]) and increase the ultrasonic velocity. The opposite effect, negative S and Δa_r is observed in the synthesis of ATP accompanied by conversion of phosphocreatine to creatine, which could be explained by the lower level of hydration of products of this reaction, when compared with the reactants.

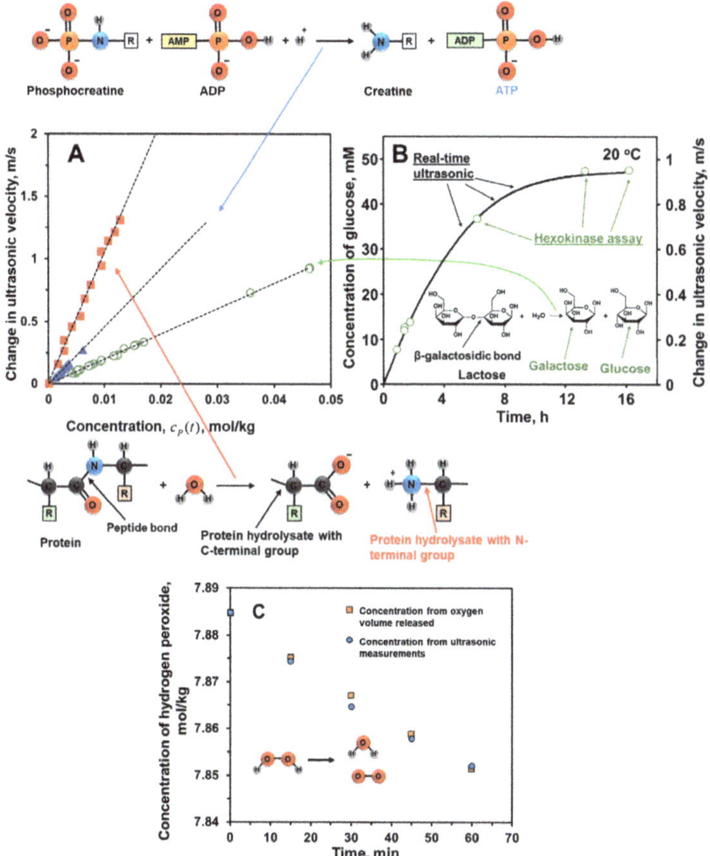

Figure 4. Application of discontinuous techniques (Method 2) for measurements or verification of Δa_r.

Figure 4A. *Change of ultrasonic velocity, $u(t) - u^0$, with concentration $c_P(t)$ in different reactions: hydrolysis of peptide bonds, hydrolysis of lactose and synthesis of ATP.*

Red squares: Reaction (R5), hydrolysis of β-lactoglobulin (0.543 mM) catalyzed by α-chymotrypsin from bovine pancreas (0.1 g/L) in 0.1 M phosphate buffer at pH 7.8, 25 °C. The ultrasonic velocity was measured at frequency 15.5 MHz. $c_P(t)$ represents the concentration of peptide bonds hydrolyzed, as determined by 2,4,6-Trinitrobenzenesulfonic acid (TNBS) method following Adler-Nissen protocol [34]. Blue triangles: Reaction (R6), synthesis of ATP (adenosine 5′-triphosphate) from ADP (adenosine 5′-diphosphate, 3.6 mM, 6.2 mM and 9.0 mM) and phosphocreatine (14.8 mM) by creatine phosphokinase from rabbit muscle (0.02 g/L) in an aqueous buffer solution (50 mM gly-gly buffer, 0.02% BSA, 5 mM Mg Acetate) at pH 7.4 and 30 °C. $c_P(t)$ represents the concentration of creatine (and ATP) formed, as determined with colorimetric hexokinase assay by monitoring the production of NADPH measured at 340 nm (hexokinase/glucose-6-phosphate dehydrogenase assay, Megazyme Int. Ltd., Bray, Ireland). For this reaction $u(t) - u^0 < 0$, however, for illustration purposes the sign of $u(t) - u^0$ was changed to positive. The ultrasonic velocity was measured at frequency 9.0 MHz.

Green circles: Reaction (R4), hydrolysis of lactose (47.9, 19, 9.83 and 5.65 mM) catalyzed by β-galactosidase (*Escherichia coli*, 369 U/g) at 20 °C in 50 mM phosphate buffer (10 mM MgCl$_2$, 10 mM mercaptoethanol) at pH 7.3. $c_P(t)$ represents the concentration of β-galactosidic bonds hydrolyzed (glucose formed) as determined with colorimetric hexokinase assay by monitoring the production of NADPH measured at 340 nm (Hexokinase/Glucose-6-phosphate dehydrogenase assay, Megazyme Int. Ltd.) [8]. The ultrasonic velocity was measured in the frequency range 2 to 16 MHz, no effect of frequency was observed.

Ultrasonic measurements were performed with HR-US 102P ultrasonic spectrometer (Sonas Technologies, Dublin, Ireland).

Figure 4B. *Comparison of ultrasonic and hexokinase assay profiles of hydrolysis of lactose by β-galactosidase in aqueous solution.*

Reaction (R4), hydrolysis of lactose (47.9 mM) by β-galactosidase (*Escherichia coli*, 369 U/g) at 20 °C in 50 mM phosphate buffer (10 mM MgCl$_2$, 10 mM mercaptoethanol) at pH 7.3. The concentration of hydrolyzed β-galactosidic bonds of lactose was calculated from the ultrasonic velocity profile (HR-US 102 ultrasonic spectrometer) using Equation (5) and Δa_r = 0.0144 kg/mol [8]. The circles represent the concentration of glucose determined with hexokinase assay in the aliquots taken from the reaction vessel at different reaction times. Frequency 5 MHz (no significant frequency dependence detected within the analyzed range 2 to 15 MHz). Adapted with permission from Reference [8]. Copyright © 2016 American Chemical Society.

Figure 4C. *Comparison of ultrasonic and volumetric (oxygen gas released) monitoring of decomposition of hydrogen peroxide in water at 25.0 °C catalyzed by an iron surface.*

Reaction (R7). Frequency 2 to 12 MHz, HR-US 102SS ultrasonic spectrometer. The samples were taken at different reaction time from the reaction vessel connected to a Hempel gas burette. No significant frequency dependence was detected within the analyzed frequency range. The ultrasonic velocity was converted to the concentration of hydrogen peroxide in the mixture using equation equivalent to Equation (9) as described earlier [5].

For reactions progressed in buffers and accompanied by a release/consumption of H^+ ions (e.g., $-NH_3^+ / - NH_2$ end groups of polypeptides in hydrolysis of proteins and $-OH / - O^-$ groups of phosphates in synthesis of ATP) the hydration effects of protonation/deprotonation of molecules of buffer also contribute to Δa_r. This contribution can be quantified through measuring the change of ultrasonic velocity caused by the protonation/deprotonation of buffer (acid or base titration of buffer). Such data can be used for recalculations of value of Δa_r obtained in a particular buffer into the value for another buffer.

Figure 4A shows linear dependence of the change in ultrasonic velocity in solutions of β-lactoglobulin with concentration of peptide bonds hydrolyzed. This agrees well with the kinetic mechanism of hydrolysis of globular proteins (including β-lactoglobulin) catalyzed by proteases. The hydrolysis of these proteins follows the so-called 'one-by-one' mechanism, in which the first step, demasking, represents the hydrolysis of one or several peptide bonds on protein surface that allow an enzyme to attack other bonds. Demasking is a limiting (slow) stage of the reaction. When the protein globule is 'opened' by demasking, the protein is hydrolyzed at a significantly higher rate. Therefore, polypeptides with intermediate degree of hydrolysis are nearly absent in the reaction mixture [36,37].

Figure 4B,C show good correlation between the ultrasonic reaction progress curves (parameter Δa_r obtained by Method 1) and the progress curves obtained with a discontinuous technique for hydrolysis of lactose and decomposition of hydrogen peroxide [5,8]. The results of the application of Method 2 are outlined in Table 1.

2.4.4. Method 3

For reactions with constant (over the reaction) Δa_r achieving saturation (plateau on the reaction progress curve), the value of Δa_r can be obtained from the values of $u(t) - u^0$ and $c_P(t)$ at saturation ($t = \infty$) according to the following relationship derived from Equation (5):

$$\Delta a_r = \frac{u(t = \infty) - u^0}{u_0 c_P(t = \infty)} \tag{14}$$

An example of this is the synthesis of ATP (Reaction (R6)) illustrated in Figure 8A. For the conditions at which the reaction was carried out, the equilibrium of the reaction is shifted to the products [38]. Therefore, at saturation $c_P(t = \infty)$ represents the known concentration of the reactant (ADP) at time zero. According to Table 1 obtained by Method 3 $\Delta a_r = -0.0310 \pm 0.0003$ kg/mol for this reaction is in good agreement with the result of Method 2, $\Delta a_r = -0.0306 \pm 0.0009$ kg/mol.

2.4.5. Method 4

For dilute mixtures in the absence of specific interactions between the components of reactants and products, the parameters a_R and a_P or Δa_r can be estimated from the available databases of ultrasonic or thermodynamic characteristics of the relevant atomic groups of substrates and products (see as example discussion in [25]). This could be utilized in development of 'libraries' of concentration increments of ultrasonic velocity of different reactions.

3. Monitoring of Chemical Reactions with Ultrasonic Attenuation

Ultrasonic attenuation, α, which represents the energy losses in compressions and decompressions of medium in ultrasonic wave, contains several contributions: classical, α_{class}, scattering, α_{scatt}, and intrinsic, α_{in}. In non-concentrated mixtures with limited level of attenuation these contributions are expected to be additive:

$$\alpha = \alpha_{class} + \alpha_{in} + \alpha_{scatt} \tag{15}$$

In non-concentrated dispersions, α_{in} is an additive sum of the intrinsic attenuation in the particles and in the continuous medium weighted by their volume fractions.

The classical contribution, α_{class}, represents the energy loses in the shear 'portion' of the longitudinal deformation of the medium (reaction mixture) in the wave, α_η, and also the energy losses caused by the heat flow between the parts of ultrasonic wave of different temperature, α_{TH}:

$$\alpha_{class} = \alpha_\eta + \alpha_{TH}$$
$$\alpha_\eta = \frac{8\pi^2}{3} \frac{\eta}{\rho u^3} f^2; \ \alpha_{TH} = 2\pi^2 \frac{T\rho\kappa e^2}{u c_P^2} f^2 \tag{16}$$

where η is the viscosity (Pa s), ρ is the density (kg/m^3), u is the ultrasonic velocity, $e = -\frac{1}{m}\left(\frac{\partial V}{\partial T}\right)_P$ is the specific thermal expansion (m^3/(K kg)) of the medium, V is the volume of the medium and m is its mass (kg), c_P is the specific thermal capacity (J/(K kg)), κ is the thermal conductivity (W/(m K)) of the medium and T is the temperature in the medium (K) [39]. The physical properties of the medium (viscosity, specific thermal expansion, specific thermal capacity and the thermal conductivity) in the above relationships represent the properties corresponding to the ultrasonic frequencies (MHz frequency range), which sometimes may be different to those measured at low frequencies or at thermodynamic limit (zero frequency).

In non-concentrated dispersions, the intrinsic contribution to ultrasonic attenuation, α_{in}, is an additive sum of the contribution of the structural relaxation in solvent/mixture [40] and (if it exists) of the relaxation processes in fast chemical reactions. If components of the analyzed reaction are involved in a relaxation process, which time scale is comparable with the period of oscillations in the ultrasonic wave, perturbations of the equilibrium in the process caused by the oscillating pressure and temperature in the ultrasonic wave produce a relaxation contribution to the ultrasonic attenuation, α_{rel}, and to the ultrasonic velocity, u_{rel}. For a relaxation process described by Reaction (R1) and characterized by the relaxation time τ, or by the relaxation frequency $f_{rel} = \frac{1}{2\pi\tau}$, the contribution to attenuation caused by this process in dilute solutions is given by Equations (A11) and (A12) of Appendix A.2. According to these equations the dependencies of the relaxation contributions $\frac{\alpha_{rel}}{f^2}$ and u_{rel} on the frequency have an S-shape profile with plateaus at low ($f \ll f_{rel}$) and high ($f \gg f_{rel}$) frequencies and half transition point at $f = f_{rel}$. At frequencies close to f_{rel}, the absolute value of both contributions decreases with frequency and vanishes at frequencies far above f_{rel}. At these high frequencies, the relaxation process is 'frozen' as its relaxation time is much longer than the period of oscillations of pressure and temperature in the ultrasonic wave. Measurements of the contribution of relaxation processes to ultrasonic parameters, especially to attenuation, provide additional tools for the monitoring of the chemical reaction.

The scattering contribution, α_{scatt}, is originated by non-homogeneity of the medium. It is discussed Appendix A.3. For particles of micron and submicron size at frequencies below 100 MHz (i.e., the wavelength of ultrasound is much longer than the particle radius), the incident ultrasonic wave is 'scattered' into the thermal and the viscous waves, originated at the border between the particle and the continuous medium. The thermal wave is produced by the oscillations of temperature gradient between the particle and the surrounding continuous medium, and the shear wave by the relative oscillatory motion of particles and of the continuous medium. In this case the ultrasonic velocity, u, and the ultrasonic attenuation, α, in a mixture consisting of the continuous medium (1) and the particles (2) of radius r can be obtained as the real (Re) and imaginary (Im) parts of the complex propagation constant K of the mixture outlined by Equation (A13) of Appendix A.3. The classical and intrinsic contributions to ultrasonic attenuation and velocity are included in the parameters α_1, α_2, u_1 and u_2 of this equation. According to Equations, for dispersions of solid particles of micron and submicron size, the dependence of the scattering contributions α_{scatt} measured at ultrasonic frequencies on the size of the particle has a 'bell' shape profile. The 'bell' has two, normally unresolved, maxima at the particle sizes close to (comparable with) the wavelength of the shear wave in the continuous medium and with the wavelength of the thermal wave in the particle and in the continuous medium (examples of the profiles are given in references [29,31,41]). The scattering contribution to ultrasonic velocity produces an increase of velocity with size following the S-shape profile with plateaus at small and large sizes and the transition sizes around the maxima of ultrasonic attenuation. Equations (A13) can be used in calculations of particle size in dispersions and its change during the analyzed chemical reactions.

4. Measuring Procedures

The commercially available HR-US spectrometers are comprised of at least two identical 'fill-in' or 'flow-through' ultrasonic cells (sample compartments), which volume ranges from 0.03 to several mL. Precision measurements of reaction progress often require the differential measuring regime of HR-US spectrometers. In this case, one of the cells is used as a measuring cell and the second as a reference cell. The ultrasonic parameters of media in both cells are monitored simultaneously during the reaction analysis. The measuring cell is filled with a medium close in composition to the reaction mixture. For example, for precision profiling of enzyme reactions in milks, milk can be used as a reference liquid, while for reactions carried out in non-concentrated aqueous solutions water can be placed in the reference cell. The measuring procedures are outlined below.

4.1. Procedure 1

Ultrasonic monitoring of the progress of analyzed reactions is carried out directly in the ultrasonic cells of HR-US spectrometers. The cells are preset to a required temperature using embedded temperature controller. The measuring cell is filled with the solution/dispersion containing the reactant and the reference cell (if differential regime is required) with an appropriate medium. The measurements of ultrasonic parameters are initiated, and the catalyst/enzyme is added to the measuring cell. The mixture is stirred using an embedded or externally controlled stirrer. The measurements of the ultrasonic velocity and attenuation are constantly performed during the reaction.

4.2. Procedure 2

The reference cell (if differential regime is required) is filled with an appropriate medium. The reaction is activated in a container by mixing of the reactants and catalyst/enzyme. The mixture is placed in the measuring ultrasonic measuring cell preset to a required temperature and the measurements are performed as in Method 1. A disadvantage of this method is the time interval (up to several minutes) required for thermal equilibration in the sample placed in the measuring cell, during which the evolution of ultrasonic velocity is affected by the changing temperature. This time interval can be shortened by bringing the temperature of reactants, catalyst and the mixing container to the temperature of the ultrasonic cell prior the mixing.

4.3. Procedure 3. Measurements in Flow

The measuring flow-through cell of the HR-US spectrometer and the external reaction container (reactor) are connected with tubing. The reference cell (if differential regime is required) is filled with an appropriate medium. The reaction is activated in the reactor and the reaction mixture is constantly pumped through the ultrasonic cell and returns back to the reactor. The ultrasonic measurements are performed in flow. The volume of the commercially available flow-through cells ranges from 0.03 to several mL.

HR-US spectrometers allow simultaneous monitoring of chemical reactions at several frequencies [23]. Although the ultrasonic velocity profiles for a significant number of reactions in solutions do not depend on frequency, this is not always the case, especially if the reactants or products are involved in a relaxation process contributing to ultrasonic velocity (see Section 3). A representative example is the hydrolysis of proteins in phosphate buffer outlined below. In this case the multifrequency measurements allowed evaluation of the effects of frequency on the measured evolution of ultrasonic velocity during the reaction, as well as provided additional information on the reaction mechanism. The scattering contributions to ultrasonic parameters as well as reactions accompanied by the formation of gel could also produce a dependence of ultrasonic velocity on frequency, which can be quantified using multifrequency measurements. In addition to this, the ultrasonic particle sizing normally requires measurements of the attenuation at several frequencies. Also, the multifrequency measurements can be utilized for assessments of the formations of non-homogeneities during catalyzed reactions, caused by phase separation, aggregation, etc., which have a pronounced effect on the profiles of velocity and attenuation measured at different frequencies.

5. Add-On Capabilities

HR-US spectrometers are equipped with precision temperature controllers, which can be utilized for the monitoring of reactions during preprogramed temperature profiles. This can be applied for assessment of the effects of temperature and various temperature profiles on the reaction progress. In this case the reaction in the ultrasonic cell is started (Procedure 1 or Procedure 2) simultaneously with the activation of the desired temperature profile during which the ultrasonic velocity and attenuation are monitored in real time.

Another useful capability includes the application of titration accessories for changing the concentration of reactants, or catalysts/enzyme, or reaction inhibitors during the reaction. In this case the reaction in the ultrasonic cell is started simultaneously with the activation of the desired concentration (titration) profile during which the ultrasonic velocity and attenuation are monitored in real time. If required, the contribution of the titrant to ultrasonic parameters can be measured separately and subtracted from the reaction curves.

6. Examples of Ultrasonic Reaction Progress Curves

This chapter discusses examples of the application of the outlined methodology of ultrasonic real-time monitoring of the evolution of concentrations of reactants and products in various reactions (reaction progress curves). The majority of the reactions were catalyzed by enzymes, however, catalysis by metal surfaces is also illustrated. The discussed reactions were carried out in aqueous solutions, in the continuous phase of complex dispersions (milks) and in nano bioreactors (nanodroplets of water in w/o microemulsion).

6.1. Hydrolysis of Maltodextrin

Figure 5 shows the real-time ultrasonic profile of hydrolysis of maltodextrin, a mixture of linear saccharides, G_n, composed mainly of 7 and 8 glucose units with average degree of polymerization 7.55 catalyzed by α-amylase (*Bacillus amyloliqu efaciens*). α-amylases are enzymes that hydrolyze the α-(1-4) linkages of such polymers as starch, amylose, amylopectin, glycogen, and dextrins [42–44]. The hydrolysis can be represented as:

$$G_{k+l} + H_2O \rightarrow G_k + G_l \tag{R3}$$

where α-(1-4) glycosidic bonds $(-O-)$ of G_{k+l} are hydrolyzed, producing oligomers of glucose with lower degree of polymerization, G_k and G_l.

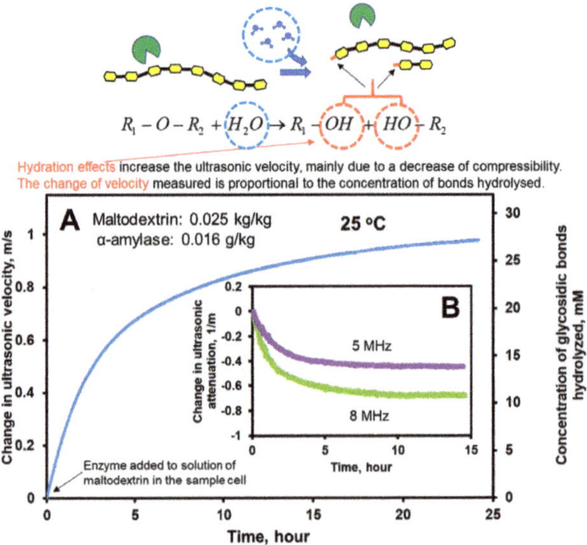

Figure 5. Real-time ultrasonic profile of hydrolysis of maltodextrin catalyzed by α-amylase. Monitoring of Reaction (R3) of hydrolysis of maltodextrin (average degree of polymerization 7.55) catalyzed by α-amylase from Bacillus sp. in 10 mM phosphate buffer at pH 6.94 at 25 °C.

Figure 5A. *Concentration profile of α-glycosidic bonds hydrolyzed.*

Calculated from the measured increase of ultrasonic velocity (HR-US 102SS spectrometer) using Equation (10) and the concentration increment of ultrasonic velocity of hydrolysis $\Delta a_r = 0.024$ kg/mol, determined by Method 1 as shown in Figure 3. Frequency 8 MHz. No effect of frequency on velocity profile was detected in the analyzed frequency range 5 to 8 MHz.

Figure 5B. *Change in ultrasonic attenuation during hydrolysis of maltodextrin catalyzed by α-amylase.*

Frequencies 5 and 8 MHz.

The reaction was started by addition of a small amount (several μL) of concentrated solution of the enzyme to the measuring ultrasonic cell of HR-US 102SS ultrasonic spectrometer preloaded with 1.4 mL of 0.025 kg/kg of maltodextrin in 10 mM phosphate buffer at pH 6.94 at 25 °C as described by Procedure 1. The concentration of the enzyme after its addition was 0.016 g/kg. The observed change in ultrasonic velocity was recalculated into the concentration (molarity) of α-glycosidic bonds hydrolyzed by applying Equations (7) and (10) and $\Delta a_r = 0.0240$ kg/mol, determined by Method 1 (Figure 3) using Equation (12), where the term a_w was neglected. The measurements were performed at frequencies 5 and 8 MHz. The ultrasonic velocity profiles at both frequencies coincide with each other within the thickness of the line on Figure 5.

The average degree of polymerization, $\overline{DP}$, and the molar mass, $\overline{M}$, were calculated according to Equation (11) and plotted in Figure 11. The degree of polymerization and the average molar mass decrease during the reaction from 7.55 and 1.224 kDa respectively to 3.31 and 0.537 kDa at reaction time of 1000 min. It is less than half of the degree of polymerization for the unhydrolyzed maltodextrin. This level of polymerization is in good agreement with the previous results on the mechanism of hydrolysis of oligosaccharides by α-amylase from *Bacillus amyloliqu efaciens* [42–44]. They demonstrated that the α-amylase cannot hydrolyze short oligomers, such as G_5, G_4, G_3 or G_2 and the hydrolysis of oligomers with 6 units, G_6, proceeds very slowly to $G_5 + G_1$ [43], while oligomers with 8 units, G_8 are mainly degraded to $G_6 + G_2$ and $G_5 + G_3$, and oligomers with 7 units, G_7 produce $G_6 + G_1$ and $G_5 + G_2$.

Overall the results presented in Figure 5 and Figure 11 (discussed below) demonstrate the capability of the HR-US technique for precision monitoring of reaction progress in the hydrolysis of oligosaccharides, which includes the real-time profiling of the average degree of polymerization and the molar mass of these molecules.

6.2. Hydrolysis of Lactose

Figure 6 illustrates the ultrasonic profiles of hydrolysis of lactose (a milk sugar) by enzyme β-galactosidase (*Kluyveromyces lactis*) in infant milk (Cow & Gate First Infant milk, 0.198 mol/kg or 6.8% kg/kg of lactose). The hydrolysis is described as:

$$Gal - O - R_2 + H_2O \rightarrow Gal - OH + HO - R_2 \tag{R4}$$

where *Gal* and R_2 represent the galactose and glucose moieties, and $-O-$ represents β-galactosidic linkage. The hydrolysis catalyzed by the β-galactosidase is accompanied by the production of galacto-oligosaccharides, GOS, which are prebiotics, described as β-linked chains of galactose units usually with terminal glucose. The term GOS includes the disaccharide galactose−galactose, however, excludes lactose. The majority of GOS produced by *Kluyveromyces lactis* β-galactosidase are di and tri-saccharides [8,45]. In the case of GOS R_2 represents galactose, or glucose, or disaccharide moieties of galactose-glucose or galactose-galactose. The GOS are produced at intermediate stages of the hydrolysis and are subsequently hydrolyzed. Small amount (0.7% kg/kg) of GOS are also present in infant milk as health benefit additives. For this type of hydrolysis, with multiple reactants and intermediate products, the reaction progress can be described by the evolution the concentration of β-galactosidic bonds hydrolyzed in the mixture (left Y scale on Figure 6A,B, molarity units). This parameter represents the difference between the concentration of the linkages removed from (hydrolysis of lactose and GOS) and added to (synthesis of GOS) the mixture. The presented in Figure 6 concentration of β-galactosidic bonds hydrolyzed was calculated using Equations (7) and (10). The concentration increment of ultrasonic velocity of hydrolysis of the $-O-$ linkage, Δa_r, was measured directly in milk using Method 1 (Figure 2A,C) and verified in aqueous solution using Method 2 (hexokinase assay, Figure 4) [8]. The reaction extent (supplementary Y axis) represents the ratio of the concentration of β-galactosidic bonds in the milk hydrolyzed at time t, $c_{bh}(t)$, to the amount of bonds at time zero hydrolysable by the enzyme, $c_b{}^0$ (0.198 mol/kg of the bonds of lactose and 0.025 mol/kg of the bonds of galactooligosaccharides). The ultrasonic measurements were collected in the frequency range 2 to 15 MHz. As no significant effects of frequency on the ultrasonic velocity profiles were observed, only the data at 5 MHz are presented. Within the limits of experimental uncertainty (1% of attenuation α) no significant changes of ultrasonic attenuation were observed during the hydrolysis, thus, indicating an absence of effects of hydrolysis on microstructural characteristics of the dispersion.

The illustrated reaction progress curves allow assessment of the impact of temperature and concentration of enzyme on performance of β-galactosidase in formulations for the reduction of levels of lactose in infant milks [8]. These formulations are added to infant's milk bottles prior to baby feeding in order to overcome the frequently observed intolerance to lactose, a serious issue in the healthy development of infants.

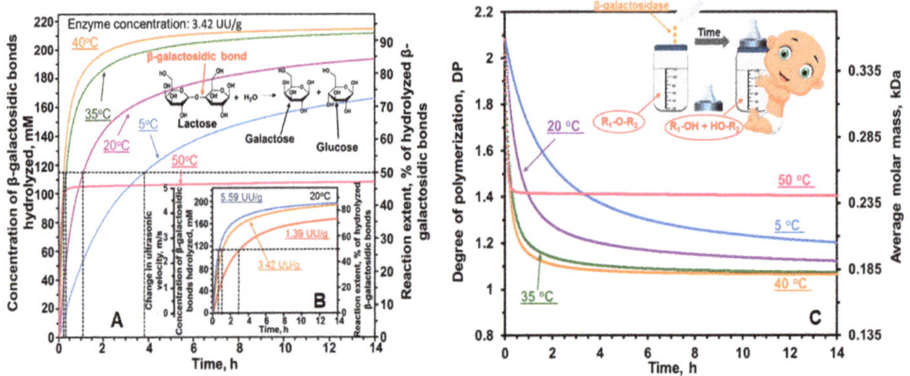

Figure 6. Real-time ultrasonic profiles of hydrolysis of lactose in infant milk by β-galactosidase.

Figure 6A. *Monitoring of Reaction (R4) of hydrolysis of lactose by enzyme in infant milk.*

β-galactosidase from Kluyveromyces lactis, Cow & Gate First Infant milk, 5 °C, 20 °C, 35 °C, 40 °C and 50 °C, enzyme concentration 3.42 UU/g.

Figure 6B. *Hydrolysis at 20 °C and enzyme concentrations 5.59, 3.42 and 1.39 UU/g.*

The concentration (molarity) profiles of β-galactosidic bonds hydrolyzed by β-galactosidase were calculated from the measured increase in ultrasonic velocity (HR-US 102PT) using Equations (7) and (10), and the concentration increment of ultrasonic velocity of hydrolysis calculated from the relationship obtained earlier [8]: $\Delta a_r(T) = 0.01812 - 2.0034 \times 10^{-4}T + 1.4713 \times 10^{-6}T^2$ kg/mol. One ultrasonic activity unit, UU, represents the amount of enzyme, which hydrolyses 1 μmol of β-galactosidic bonds in Cow & Gate First Infant milk per min at 20 °C and reaction extent equal to zero [8]. The ultrasonic measurements were collected in the frequency range 2 to 15 MHz. As no significant effects of frequency on the ultrasonic reaction profiles were observed only the data at 5 MHz are presented. Adapted with permission from Reference [8]. Copyright © 2016 American Chemical Society.

Figure 6C. *Average degree of polymerisation and molar mass of the oligosaccharides (lactose +GOS) present in the infant milk during the hydrolysis of lactose by β-galactosidase.*

Calculated from the data of Figure 6A using Equation (11). The initial degree of polymerization was obtained from UPLC-HILIC-FLD profiles of the milk published earlier [8]).

According to Figure 6A the time required for the reduction in concentration of the β-galactosidic bonds to a particular level depends significantly on the temperature of the milk. A reduction of 50% is achieved at 14 min for 40 °C and 226 min for 5 °C. At 50 °C the reaction extent levels off at 45.6% [8]. This demonstrates the fast deactivation of the enzyme at 50 °C, which is in general agreement with the results obtained for this enzyme in aqueous solutions [46]. The outlined dependence of the enzyme activity on temperature represents one of the key challenges for practical applications of β-galactosidase in household environments.

The inset of Figure 6B shows that the time required for a particular level of hydrolysis at 20 °C is inversely proportional to enzyme concentration.

Figure 6C shows the evolution of the average degree of polymerization and the average molar mass of the oligosaccharides (lactose + GOS) present in the infant milk during hydrolysis of lactose calculated according to Equation (11). The degree of polymerization at time zero was obtained from the concentrations of lactose and added GOS (calculated from Table 1 and UPLC-HILIC-FLD profiles in Caras Altas et al. [8]) is 2.08. At temperature 40 °C and reaction time 14 h, the degree of polymerization is approximately 1.06, indicating almost fully complete hydrolysis to monomers glucose and galactose, which is in good agreement with previously obtained conclusions on the high level of hydrolysis at this temperature [45,47]. At 50 °C the degree of polymerization is approximately 1.4 at 14 h of reaction, which is explained by the fast deactivation of the enzyme at this temperature.

6.3. Hydrolysis of Proteins

Figure 7 shows real-time ultrasonic profiles of hydrolysis of globular protein β-lactoglobulin (from bovine milk) at different concentrations (0.547 and 0.273 mmol/kg) catalyzed by α-chymotrypsin in 0.1 M potassium phosphate buffer at pH 7.8 at 25 °C. The hydrolysis is described by the following reaction:

$$P_1 - C(O) - NH - P_2 + H_2O \rightarrow P_1 - COO^- + -NH_3^+ - P_2 \tag{R5}$$

where $P_1 - COO^-$ and $-NH_3^+ - P_2$ represent the protein hydrolysates with C-terminal group and protein hydrolysates with N-terminal group, respectively, and $C(O) - NH$ represents the peptide bond.

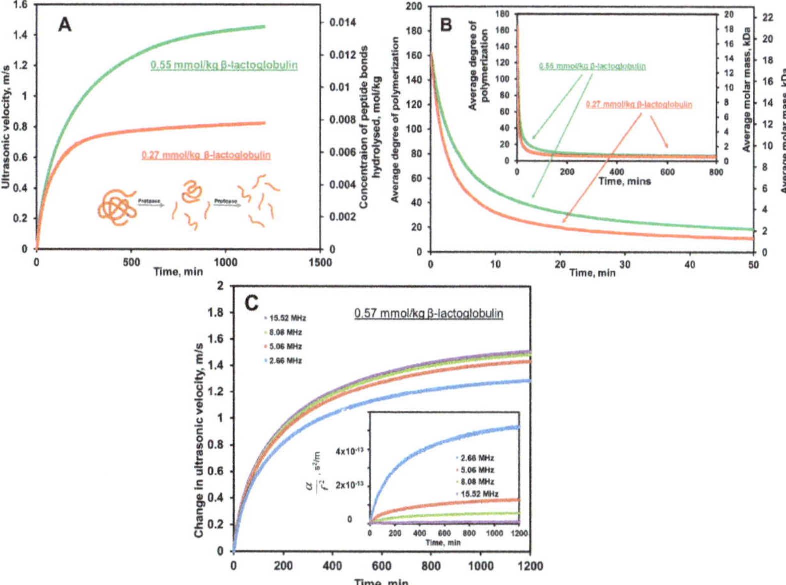

Figure 7. Real-time ultrasonic profiles of hydrolysis of protein β-lactoglobulin by proteolytic enzyme α-chymotrypsin. Reaction (R5), hydrolysis of β-lactoglobulin, in 0.1 M potassium phosphate buffer at pH 7.8 at 25 °C catalyzed by protease α-chymotrypsin from bovine pancreas 0.0021 kg/kg.

Figure 7A. *Time profiles of ultrasonic velocity at frequency 15 MHz and concentration of peptide bonds hydrolyzed.*

Concentrations of β-lactoglobulin 0.55 and 0.27 mmol/kg. The concentration of peptide bonds hydrolyzed calculated using Equation (10), and $\Delta a_r = 0.070$ kg/mol.

Figure 7B. *Real-time profiles of average degree of polymerization and molar mass.*

Calculated from the concentration of bonds hydrolyzed profile shown in Figure 7A according to Equation (11).

Figure 7C. *Time profiles for ultrasonic velocity and attenuation measured at different frequencies.*

The hydrolysis was activated by addition of several μL of concentrated solution of enzyme (α-chymotrypsin from bovine pancreas) to 1.1 mL solution of β-lactoglobulin in 0.1 M phosphate buffer at pH 7.8 preloaded into the ultrasonic cell of HR-US 102 ultrasonic spectrometer as described by Procedure 1. The concentration of enzyme in the ultrasonic cell was 0.0021 kg/kg. The ultrasonic velocity profiles were recalculated into the profiles of concentration of peptide bonds hydrolyzed, by applying Equation (10) and using $\Delta a_r = 0.070$ kg/mol determined by Method 2 (Table 1).

In contrary to the majority of reactions covered in this paper, the amplitude of the change of velocity caused by the hydrolysis of β-lactoglobulin in phosphate buffer is dependent on frequency, as illustrated by Figure 7C. The origin of this dependence is the relaxation process associated with the proton transfer between $-NH_3^+$ and $-NH_2$ terminal amino groups of the protein hydrolysates and the phosphate ions as discussed in the Section 9.2. According to Figure 7C, the amplitude of velocity rise becomes to be frequency independent above 10 MHz (approximately). This result agrees with the profiles of ultrasonic attenuation shown in the inset of Figure 7C (Section 9.2), thus, indicating the 'freezing' of the relaxation process (see Section 3 and Appendix A.2) at frequencies above 10 MHz. Therefore, the velocity profiles obtained at frequencies 15 MHz were used for determination of Δa_r utilizing Method 2 (Figure 4), and for the subsequent calculations of the concentration of peptide bonds hydrolyzed according to Equation (10) from the measured change in ultrasonic velocity.

Figure 7B shows the change of the average degree of polymerization, $\overline{DP}$, and the molar mass, $\overline{M}$, during the hydrolysis calculated from the concentration of peptide bonds hydrolyzed, $c_{bh}(t)$, according to Equation (11). The degree of polymerization at time zero, $\overline{DP}^0$, was taken as 162, which corresponds to the number of amino acid residues in β-lactoglobulin. Accordingly, the molar mass at time zero was taken as 18.4 kDa, which represents the molar mass of the 162 amino acids linked with the peptide bonds between them. At our experimental conditions (pH and ionic composition), β-lactoglobulin in solution is expected to exist in the forms of single molecules and the molecules aggregated into dimers [48]. Therefore, the average molar mass plotted in Figure 7B could be different to that measured with light or neutron scattering techniques or other methods, which probe the size of aggregated protein structures.

For concentration of 0.27 mmol/kg (0.005 kg/kg approx.) the average degree of polymerization, $\overline{DP}$, and the molar mass, $\overline{M}$, change from 162 and 18.4 kDa for unhydrolyzed protein to 5.5 and 0.64 kDa respectively at 1200 min. For concentration of 0.55 mmol/kg (0.01 kg/kg approx.) those values change to 6.2 and 0.70 kDa at 1200 min respectively.

The enzyme α-chymotrypsin cleaves peptide bonds selectively on the carboxyl-terminal side of the large hydrophobic aromatic amino acids such as tryptophan, tyrosine, phenylalanine, with high catalytic efficiency [49,50]. Apart from these amino acids, α-chymotrypsin also hydrolyses, although more slowly, the peptide bonds on the carboxyl-terminal side of the leucine and methionine [51,52]. Therefore, the primary structure of β-lactoglobulin allows hydrolysis by α-chymotrypsin of 36 peptide bonds of 162 amino acids present. This corresponds to the limiting value of the degree of polymerization of 4.4, which is lower, however, close to the values obtained at long reaction times.

Hydrolysis of proteins into protein hydrolysates catalyzed by proteases is utilized in a broad range of industries, from food to effective therapeutics, since the products of this reaction have additional nutritional and functional value. Enzymatic protein hydrolysates containing short-chain peptides with characteristic amino acid composition and defined molecular size are highly desired for specific formulations, since they possess higher solubility, heat stability and valuable bioactive properties [53,54]. These hydrolysates play important roles in various aspects of health being, such as mineral binding, immunomodulatory, antioxidative, antithrombotic, hypocholesterolemic, antihypertensive function, and others [55]. Development and production of protein hydrolysates is dependent on efficient tools for real-time monitoring of the hydrolysis of peptide bonds under different environmental conditions in bioreactors. The results discussed in this chapter illustrate the potential of application of HR-US technology for this task.

6.4. Synthesis of ATP

Figure 8 illustrates ultrasonic real-time profile of synthesis ATP by the transphosphorylation of phosphocreatine to ADP catalyzed by creatine kinase at 30 °C described by reaction:

$$ADP + phosphocreatine \rightarrow ATP + creatine \qquad (R6)$$

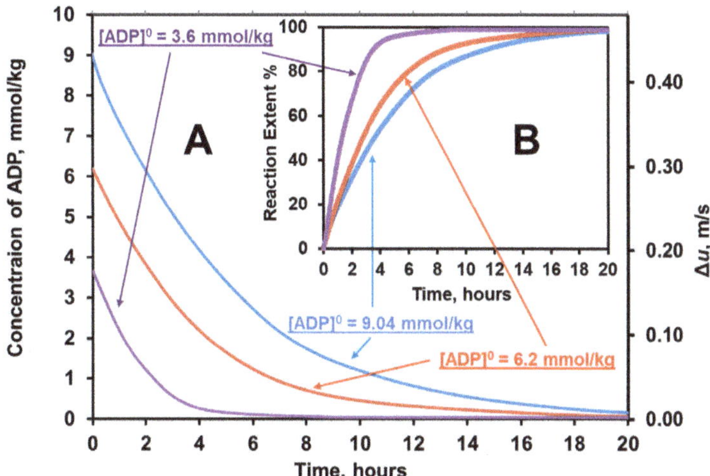

Figure 8. Real-time ultrasonic profiles of synthesis of ATP.

Figure 8A. *Ultrasonic profile of concentration of ADP during synthesis of ATP.*

Reaction (R6). Synthesis of ATP from ADP (3.6, 6.2 and 9.0 mmol/kg) accompanied by conversion of phosphocreatine to creatine by enzyme creatine phosphokinase from rabbit muscle. The reaction was performed in a buffer solution (50 mM gly-gly buffer, 0.02% BSA, 5 mM Mg Acetate at pH 7.4) at 30 °C. The concentration profile was calculated from the measured change in ultrasonic velocity at 9.0 MHz (HR-US 102 spectrometer), using Equation (5) and the concentration increment of ultrasonic velocity of hydrolysis, $\Delta a_r = 0.031$ kg/mol. For illustration purposes the decrease of ultrasonic velocity during reaction is presented as $\Delta u = u(t) - u(t = \infty)$ were $u(t = \infty)$ is the ultrasonic velocity at reaction completion.

Figure 8B. *Evolution of reaction extent in synthesis of ATP.*

Calculated from the ultrasonic velocity profile using Equation (5).

The reaction was activated by mixing of enzyme and the solutions of ADP and phosphocreatine in 50 mM gly-gly buffer, 0.02% BSA, 5 mM Mg acetate at pH 7.4 and loaded into the measuring ultrasonic cell of the HR-US 102 ultrasonic spectrometer pre-equilibrated at temperature 30 °C, following Procedure 2. The reaction was carried out at three different concentrations of ADP,

3.6 mmol/kg, 6.2 mmol/kg and 9.0 mmol/kg, and at same concentration of phosphocreatine, 15 mmol/kg. The concentration of the enzyme was 0.18 µg/mL (185 U/mg). The ultrasonic velocity profiles were recalculated into the profiles of concentration of ADP and the reaction extent by applying Equation (5) and using $\Delta a_r = -0.0310$ kg/mol determined by Method 3 and verified by Method 2 (Table 1). The concentration of ATP synthesized during the reaction at time t is equal to the difference of the initial concentration of ADP and the concentration of ADP at time t. According to Figure 8B, which represents the reaction extent, the reaction reaches completion (reaction extent = 100%) after approximately 10 h when the lowest concentration of ADP (3.6 mmol/kg) is used.

The enzyme creatine kinase occupies an important place in the regulation of muscle bioenergetics by providing rapid replenishment of the concentration of ATP from the reservoir of phosphagens [56,57]. A variety of methods have been applied for monitoring this reaction, including direct and indirect colorimetric assays. The major benefits of application of HR-US technique is the ability to monitor it precisely during the whole course of the reaction, without optical markers and/or secondary reactions, and in various media including opaque mixtures.

6.5. Decomposition of Hydrogen Peroxide Catalyzed by Metal Surfaces

Figure 9 illustrates real-time ultrasonic monitoring of the decomposition of hydrogen peroxide at 25.0 °C catalyzed by iron surface at two concentrations of hydrogen peroxide (2.81 mol/kg for the main frame and 0.0743 mol/kg for the inset) [5]. Hydrogen peroxide is used in a variety of applications, such as wastewater treatment, preparation of semiconductor materials and cleaning liquids for printed circuit boards. The decomposition of hydrogen peroxide dissolved in water can be presented as:

$$H_2O_2 \rightarrow \frac{1}{2}O_2 + H_2O \tag{R7}$$

Prior to the measurements, water was saturated with nitrogen at the measuring temperature to provide the consistency of its composition, and to prevent solubilization of CO_2 and formation of carbonic acid. After placing the aqueous solution of hydrogen peroxide into the temperature controlled reaction container (100 mL) the reaction was activated by adding the catalyst, bare iron wire (2×50 cm, 1 mm diameter, purity 99%+) precleaned with hydrochloric acid. The reaction vessel was connected with the measuring cell of HR-US 102SS ultrasonic spectrometer using a flexible tubing passing a peristaltic pump, as described in Procedure 3, and ultrasonic measurements were performed in flow-through regime [5]. The ultrasonic velocity profile during the reaction is given as the change of the relative ultrasonic velocity, which represents the difference between the ultrasonic velocity in the solution and in pure water. As no significant dependence of ultrasonic velocity on frequency within the frequency range 2 to 12 MHz was observed, only the data collected at 12 MHz are represented in the figure.

As the solubility of oxygen in water is very low, most of the oxygen produced in this reaction is released from the solution, with the exception of the initial stages (first minutes) when the produced oxygen substitutes nitrogen in the solution. The change in ultrasonic velocity in water caused by this substitution is at a level of 0.002 m/s and can be neglected when compared with the observed changes of ultrasonic velocity in the analyzed reaction [5]. Therefore, when considering the concentrations of reactants and products in the solution the reaction can be presented as: $H_2O_2 \rightarrow H_2O$. Consequently, the concentration increment of ultrasonic velocity of the reaction, Δa_r, can be expressed through concentration increments of ultrasonic velocity of water, a_w, and hydrogen peroxide, $a_{H_2O_2}$: $\Delta a_r = a_w - a_{H_2O_2}$. As a_w represents the effect of addition of water to water on ultrasonic velocity it is equal to zero. According to Figure 2B $a_{H_2O_2} = a_{0\,H_2O_2} + a_{1\,H_2O_2}c_{H_2O_2}$, where $a_{0\,H_2O_2}$ and $a_{1\,H_2O_2}$ are constants. Thus, for the Reaction (R7) $\Delta a_r = -a_{H_2O_2} = -a_{0\,H_2O_2} - a_{1\,H_2O_2}c_{H_2O_2}$ (see Reference [5] for details). Following this, the concentration profile of hydrogen peroxide (secondary Y axis in Figure 9) during the reaction was calculated from the ultrasonic velocity profile using Equation (9) where $\Delta a_0 = -a_{0\,H_2O_2}$ and $\Delta a_1 = -a_{1\,H_2O_2}$.

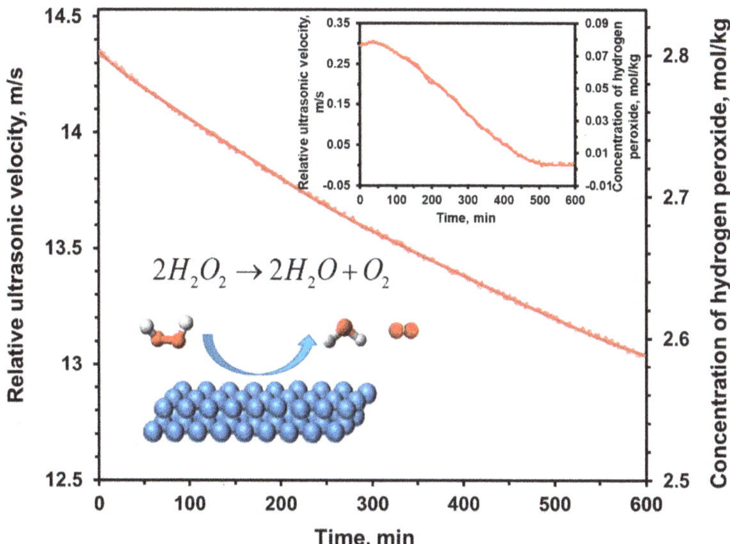

Figure 9. Real-time ultrasonic profiles of decomposition of hydrogen peroxide catalyzed by iron surface.
Concentrated (2.81 mol/kg main frame) and diluted (0.0743 mol/kg, inset) aqueous solutions of hydrogen peroxide at 25 °C. The concentration profile of hydrogen peroxide was calculated from the relative ultrasonic velocity using Equation (9), and the concentration increment of ultrasonic velocity obtained earlier [5]. Adapted from Reference [5], with permission from The Royal Society of Chemistry.

At high initial concentration of hydrogen peroxide (2.81 mol/kg at time zero) the reaction rate progressively decreases with time. It is interesting to note that the time profile of the concentration of hydrogen peroxide does not correspond to the pseudo-first order reaction often applied for catalysis on surfaces [5]. This can be explained by a complex multistep nature of the reaction [58,59], which is also suggested by the ultrasonic results at the low initial concentration of hydrogen peroxide, 0.0743 mol/kg. For this concentration, an approximately 50 min delay between the introduction of the iron surface in the solution and the beginning of decomposition of hydrogen peroxide is observed. One of the possible reasons for this delay is the 'activation' of the iron surface, which could include surface oxidation (formation of Fe^{2+} ions catalyzing the decomposition [60–66]).

6.6. Encapsulation of Substrates and Enzymes. Monitoring of Reactions in Nano-Droplets

Enzymes are conventionally used in aqueous solutions. However, water is often a poor solvent for applications in synthetic and industrial chemistry [67]. Water-in-oil microemulsions (w/o ME) provide a favorable hydrophilic environment for enzyme activity [68]. In addition to this, the confined environment inside the ME droplets affects the mobility of the enzyme atomic groups and their hydration level, which can be applied for changing/controlling the enzymatic catalytical activity [67]. Examples of applications of microemulsions for enzyme catalysis of reactions include hydrolytic and reverse hydrolytic reactions, esterifications and transesterifications, resolution of racemic amino acids, oxidation and reduction of steroids, and synthesis of phenolic and aromatic amine polymers [68].

6.6.1. Microemulsion Phase Diagrams. Optimal Conditions for Encapsulation of Enzymes and Reactants

The desired functioning of enzymes encapsulated in aqueous nanodroplets is significantly determined by the level of hydration of enzyme molecules and of the molecules of substrates and products. Therefore, successful encapsulation of enzymes and of the components of reactions involves identification of the parts of the w/o microemulsion phase diagrams where nanodroplets contain the required for the hydration of enzyme and substrate/product amount of free water. HR-US titration analysis allows for identification of these parts of microemulsion phase diagrams [31].

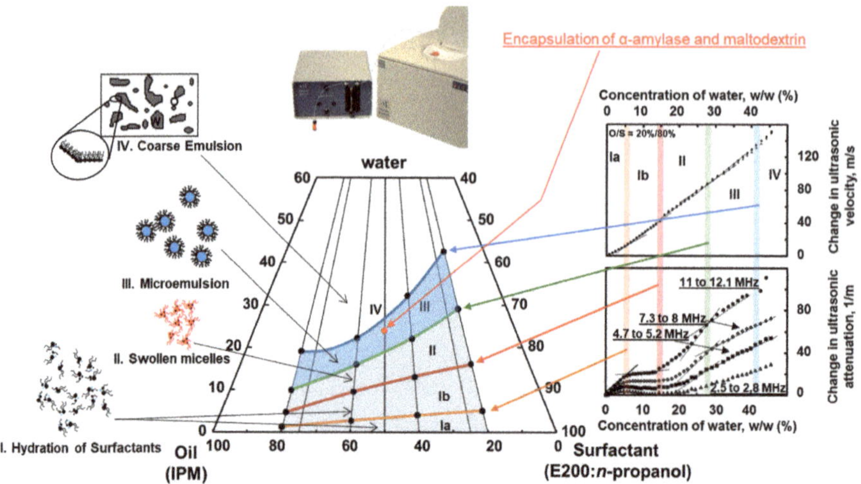

Figure 10. Ultrasonic pseudo-ternary phase diagram of IPM microemulsion.

Ultrasonic pseudo-ternary (concentration scale, % kg/kg) phase diagram of water/isopropyl myristate (IPM)/epikuron 200 (E200) and n-propanol (1:1) at 25 °C. The transition lines between different sub-phases (hydration of surfactants, swollen micelles, microemulsion and coarse emulsion) represent the 'break' points on the ultrasonic (velocity and attenuation) titration profiles of surfactant/cosurfactant and oil mixture with water. The red point shown in the phase diagram corresponds to the conditions (water (24% (kg/kg)), IPM (38% (kg/kg)), n-propanol (19% (kg/kg)), E200 (19% (kg/kg)) utilized in the hydrolysis of maltodextrin catalyzed by α-amylase in the aqueous droplets of this microemulsion system shown in Figure 11 (HR-US 102SS) ultrasonic spectrometer.

Figure 10 represents the ultrasonic phase diagram for microemulsion composed on pharmaceutically acceptable ingredients: isopropyl myristate (IPM) microemulsion (oil-IPM, surfactant-Epikuron 200 (E200)/cosurfactant-*n*-propanol (1:1), and water) at 25 °C. The diagram was obtained from titration profiles of ultrasonic velocity and attenuation, performed by automatic stepwise additions of water to 1.5 mL of a mixture of oil, surfactant and cosurfactant in the ultrasonic cell of HR-US 102SS ultrasonic spectrometer. An example of an ultrasonic titration profile in 20:80 kg/kg oil:surfactant/cosurfactant mixture is given in Figure 10. Overall, the ultrasonic titration profiles provide a number of parameters reflecting different levels of microstructural organization of the system [29,30,69]. The transitions between different states of the system are clearly identified by an abrupt change of concentration profile of ultrasonic velocity, attenuation and their frequency dependence. This includes the end of microemulsion phase represented by the blue line, above which the mixture becomes opaque (phase IV). In addition to this, the ultrasonic titration profiles demonstrate a range of transitions below the blue line, which allows distinguishing different 'sub phases', Ia, Ib, II and III. The nature of these 'sub phases' is related to the state of water and surfactant, including hydration water in the dispersed phase, hydration water in swollen reverse micelles and water in aqueous nano-size droplets surrounded by surfactant, realized in 'sub phase' III [31]. According to Figure 10 the sub-phase III shall provide a reasonable hydration level for hydrophilic enzymes, substrates and products (see more discussions in [29–31,69]).

6.6.2. HR-US Monitoring of Enzyme Catalyzed Reactions in Nanodroplets

Figure 11 compares the real-time ultrasonic profiles of hydrolysis of maltodextrin catalyzed by α-amylase from *Bacillus* sp. in aqueous solution (discussed in the Section 6.1) and in the described above w/o microemulsion consisting of oil (isoproprylmyristate, 38% kg/kg), cosurfactant (*n*-propanol, 19% kg/kg), surfactant (Epikuron 200, 19% kg/kg) and water (24% kg/kg). This composition of microemulsion was 'good' for the hydrolysis of maltodextrin, since it corresponds to the phase III of the diagram where 'free' water in macroemulsion droplets of approximately 10 nm diameter [29]. The water in the droplets shall provide a substantial hydration of the substrate (maltodextrin), of the enzyme (α-amylase) and of the products of the Reaction (R3). The concentrations of the substrate and of the enzyme in the aqueous phase of the microemulsion were close to those in the aqueous solution. The reaction in the microemulsion was started by adding of small amount (40 µL) of concentrated solution of the enzyme to the measuring ultrasonic cell of HR-US 102SS ultrasonic spectrometer preloaded with 1.4 mL of microemulsion, which aqueous phase contained 0.025 kg/kg of maltodextrin in 10 mM phosphate buffer at pH 6.94 at 25 °C according to Procedure 1.

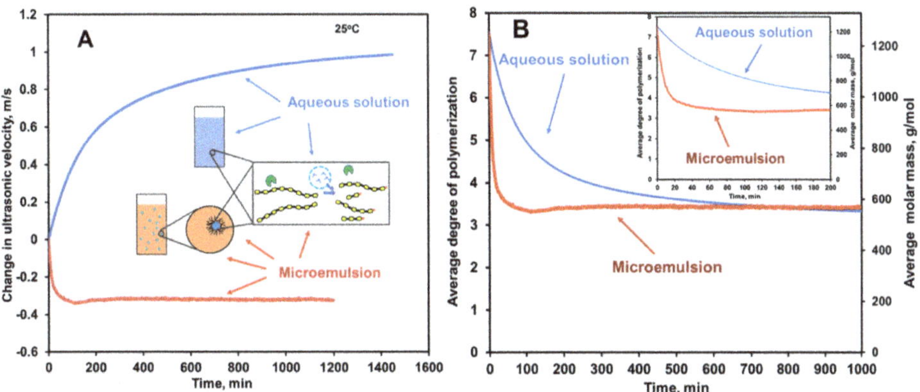

Figure 11. Real-time ultrasonic profile of hydrolysis of maltodextrin by α-amylase in aqueous solution and in nano-droplets of IPM microemulsion.
Comparison of ultrasonic profiles of hydrolysis of maltodextrin catalyzed by α-amylase (*Bacillus* sp.) in aqueous solution (conditions are outlined in the legend of Figure 5) and in microemulsion (oil (IPM, 38% kg/kg), cosurfactant (n-propanol, 19% kg/kg), surfactant (E200, 19% kg/kg) and water (24% kg/kg)).

Figure 11A. *Ultrasonic velocity profile of hydrolysis of maltodextrin catalyzed by α-amylase in aqueous solution and in microemulsion.*

Maltodextrin with concentration 0.025 kg/kg at 25 °C in 10 mM of phosphate buffer at pH 6.94. The concentrations of enzyme are 0.016 g/kg for aqueous solution and 0.015 g/kg for the aqueous phase of microemulsion system. HR-US 102SS spectrometer. Frequency 8 MHz. No effect of frequency on velocity profile was detected in the analyzed frequency range, 5 to 8 MHz for aqueous solution and 3 to 8 MHz for microemulsion. The concentrations of maltodextrin and the enzyme in microemulsion are given per kg of aqueous phase (buffer).

Figure 11B. *Real-time profiles of average degree of polymerization and of molar mass in aqueous solution and in microemulsion.*

Calculated from ultrasonic velocity profiles shown in Figure 11A according to Equation (11).

The ultrasonic velocity profiles were recalculated into the average degree of polymerization and the molar mass of oligosaccharides in the reaction mixture using Equation (11). The value of

$\Delta a_r = -0.0093$ kg/mol in the microemulsion was determined according to Method 1 and Equation (12) using the data shown in Figure 3 ($b_0 = -0.0062$ kg/mol) and the concentration increment of ultrasonic velocity of water, $a_w = 0.0031$ kg/mol measured by additions of small amounts of water to the microemulsion. For microemulsions a_w represents the slope of the plot of ultrasonic velocity vs. concentration of water illustrated in Figure 10. The negative value of Δa_r in microemulsion looks unusual, if compared with aqueous solutions. However, this can be explained by the effects of the heat exchange between the aqueous and oil phases of microemulsion during the compression cycle in the ultrasonic wave on the apparent compressibility of solutes solubilized in microemulsion nano-droplets as discussed previously [31]. The ultrasonic measurements in microemuslion were collected in the frequency range 3 to 8 MHz. As no significant effects of frequency on the ultrasonic reaction profiles were observed, only the data at 8 MHz are presented.

According to Figure 11, the reaction in microemulsion proceeds significantly faster than in solution, which could be explained by a super-activity of the enzyme, often observed in microemulsions [70–72]. This conclusion is in agreement with the previous observations of the increase in the hydrolytic activity of the α-amylase against the oligomer substrates in a presence of surfactants [42]. The average degree of polymerization and the molar mass of maltodextrin at 100 min of reaction in microemulsion are 3.3 and 0.56 kDa respectively, which is close to the values obtained in solutions at significantly longer reaction time, 1000 min, as illustrated by Figure 11B. It is interesting that in microemulsion, the decrease in ultrasonic velocity and in the degree of polymerization within 100 min of the reaction is followed by a slow increase over a longer period of time. The increase could be attributed to the previously observed alternative synthetic activities of α-amylase in restricted conditions of encapsulated state. The activities include: (1) synthesis of $-O-$ bonds between moieties of glucose (G) in maltodextrin at the initial stage of reaction forming molecules with high degree of polymerization ($G_n + G_k \rightarrow G_{n+k} + H_2O$); (2) synthesis of $-O-$ bonds between the moieties of glucose and n-propanol (P) present in microemulsion as cosurfactant and formation alkyl glucosides ($G_n + kP \rightarrow G_n P_k + H_2O$) [73,74].

7. Ultrasonic Reaction Rates and Advance Chemical Kinetics

High precision of ultrasonic measurements of concentration of substrates and products during analyzed reactions can be utilized in obtaining the reaction rates using Equation (8) and appropriate digital differentiation (concentration vs. time) procedures. This provides the detailed 'reaction rate vs. concentration of reactant/product' profiles over the whole course of reaction. Typically, the advanced modeling of kinetic mechanisms of chemical reaction is based on functional relationships between the reaction rate and the concentrations of reactants and products. The ultrasonic 'reaction rate vs. concentration of reactant/product' profiles can be applied for verification of these relationships and determination of the underlying kinetic and thermodynamic constants, including those involved in complex mechanisms of inhibition as illustrated below.

Figure 12 represents an example of the ultrasonic reaction rate vs. concentration of reactant profiles for hydrolysis of cellobiose (glucose disaccharide) at its different initial concentrations catalyzed by cellobiase (β-glucosidase from *Aspergillus niger* (Novozyme 188)), in 10 mM acetate buffer (pH 4.9) at 50 °C measured using Procedure 2 [25]. The overall reaction can be expressed as:

$$G_2 + H_2O \rightarrow 2G \tag{R8}$$

where G represents the molecule of glucose and G_2 the molecule of cellobiose. The 'reaction rate vs. concentration of reactant' profiles were calculated from the measured ultrasonic reaction curves ($\Delta a_r = 0.0126$ kg/mol) by averaging the slopes of two adjacent points in combination with the Savitzky–Golay smoothing method [25].

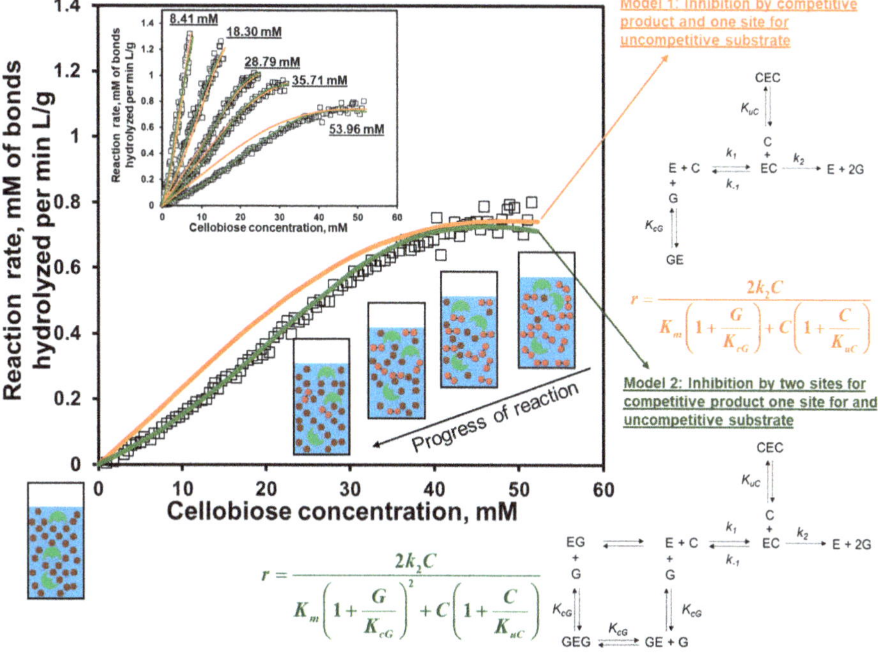

Figure 12. Ultrasonic reaction rate profile in hydrolysis of cellobiose by enzyme cellobiase. Evaluation of reaction inhibition mechanism.

Squares: Experimental reaction rates per unit of concentration (g/L) of enzyme, during hydrolysis of cellobiose, Reaction (R8).

50 °C in 10 mM acetate buffer at pH 4.9. Initial concentrations of cellobiose (curves from left to right in the inset): 8.41, 18.30, 28.79, 35.71, and 53.96 mM [25].

Lines: Fittings applied to the reaction rates for verification of two kinetic Models 1 and 2.

Orange lines: Model 1: uncompetitive substrate (cellobiose, C) and competitive product (glucose, G) inhibitions. Green lines: Model 2: uncompetitive substrate (cellobiose, C) and two sites for competitive product (glucose, G) inhibitions. Adapted with permission from Reference [25]. Copyright © 2011 Elsevier Inc.

Previous studies of kinetics of the hydrolysis of cellobiose catalyzed by β-glucosidase from *A. niger* have shown that the reaction follows uncompetitive cellobiose and competitive glucose inhibitions [25,75,76]. The proposed model of such inhibition is illustrated in Figure 12 (Model 1). According to this model, glucose binds to the free enzyme decreasing the apparent affinity of the substrate, whereas cellobiose inhibits hydrolysis by binding to the enzyme–substrate complex, thus reducing its effective concentration. The parameter $K_m = \frac{k_{-1}+k_2}{k_2}$ represents the Michaelis constant, k_1 and k_{-1} are the rate coefficients for the formation of complex EC, and k_2 is the catalytic coefficient (see Figure 12). The inhibition constants are presented by the parameters K_{cG} and K_{cG} (constants of dissociation of GE to $G + E$ and CEC to $EC + C$). The expression for the reaction rate, r, for this model [77] is given as:

$$r = \frac{2k_2 C}{K_m\left(1+\frac{G}{K_{cG}}\right)+C\left(1+\frac{C}{K_{uC}}\right)} \tag{17}$$

where C is the concentration of cellobiose and G is the concentration of glucose in the reaction mixture. Figure 12 shows the ultrasonic reaction rate profiles (squares) over the course of reaction for different initial concentrations of cellobiose (8.41, 18.30, 28.79, 35.71, and 53.96 mM) and their fitting (combined fitting of all curves) using Equation (17).

Although Model 1 describes reasonably well the profiles for low initial concentrations of cellobiose (8.41, 18.30 and 28.79), a significant deviation of this mathematical model from the experimental data can be noticed at the highest initial concentration of cellobiose (53.96 mM, main frame of Figure 12). More importantly, at a high initial concentration of cellobiose, the experimental reaction rate curve exhibits an S-shape profile that cannot be reproduced by Equation (17) [25]. Thereby, a new, updated kinetic inhibition model was proposed as illustrated in Figure 12 (Model 2). Model 2 suggests two (instead of one) binding sites for glucose on the enzyme. For simplicity, it was suggested that both sites are characterized by the same dissociation constant, K_{cG}. Additionally, it was speculated that the second site for glucose binding may be related to the known secondary, transgalactosylation, activity of the enzyme [25]. The equation for the rate representing Model 2 is given as:

$$r = \frac{2k_2 C}{K_m \left(1 + \frac{G}{K_{cG}}\right)^2 + C\left(1 + \frac{C}{K_{uC}}\right)} \tag{18}$$

Figure 12 shows the fitting (combined fitting of all curves) of the experimental data with Equation (18) at different initial concentrations of cellobiose. This fitting describes the experimental results very well and provides the kinetic and thermodynamic constants of the reaction, k_2, K_m, K_{cG} and K_{uC} [25].

8. Application of Extended Capabilities of Ultrasonic Spectroscopy

8.1. Dynamic Range

One of the important features of HR-US technique is its capability to work with solutions of low and high concentrations (e.g., dynamic range). This includes precision monitoring of changes in small fraction of reactants presented in large concentration. Applications of this feature are illustrated by the two following examples.

8.1.1. Ultrasonic Monitoring of Reverse Reactions

Figure 13A illustrates the ultrasonic reaction progress curve of hydrolysis (Reaction (R8)) of disaccharide cellobiose consisting of two molecules of glucose catalyzed by β-glucosidase from *Aspergillus niger* in 10 mM acetate buffer at pH 4.9 at 50 °C measured using Procedure 2 [25]. The parameter $\Delta a_r = 0.0126$ kg/mol was obtained by Method 1, and verified by Method 2 (hexokinase assay).

Figure 13B illustrates the ultrasonic reaction progress curve of the reverse Reaction (R8), synthesis of cellobiose from glucose, catalyzed by the same enzyme. The synthesis was initiated by adding the enzyme β-glucosidase to the solution of glucose in 10 mM acetate buffer at pH 4.9 at 50 °C using Procedure 2 [25]. At equilibrium in Reaction (R8) the concentration of cellobiose represent only small fraction of concentration of glucose. Therefore, relatively high concentrations of glucose (0.56 mol L^{-1} for the profile presented in the Figure 13B,C) were used. The presented ultrasonic velocity profile was obtained at frequency 5.3 MHz, however, no significant effect of frequency on the profiles were observed in the analyzed frequency range 2 to 20 MHz [25]. The change in the concentrations of glucose was calculated from the ultrasonic velocity profiles using Equations (5) and (7) and the parameter Δa_r corrected for small deviation from the above value caused by high concentration of glucose [25]. As can be seen from the Figure 13C less than 3% of glucose present in the solution reacted at the saturation level. In spite of this small fraction of glucose consumed by the synthesis, the reaction profile is well resolved by the HR-US technique. The obtained saturation level provides the equilibrium concentration of glucose in the solution for Reaction (R8), which allows calculations of the equilibrium constant of the reaction, K_{eq}. The equilibrium constants K_{eq} (expressed through mole fractions of cellobiose, water and glucose) obtained in this way in solutions of different concentration of glucose are shown in Figure 13D and correspond to the average value of 1.8 ± 0.1. The obtained equilibrium constants allowed the calculations of the standard Gibbs energy of reaction $\Delta G^0 = -8$ kJ/mol, which agrees with the results obtained by other techniques [25].

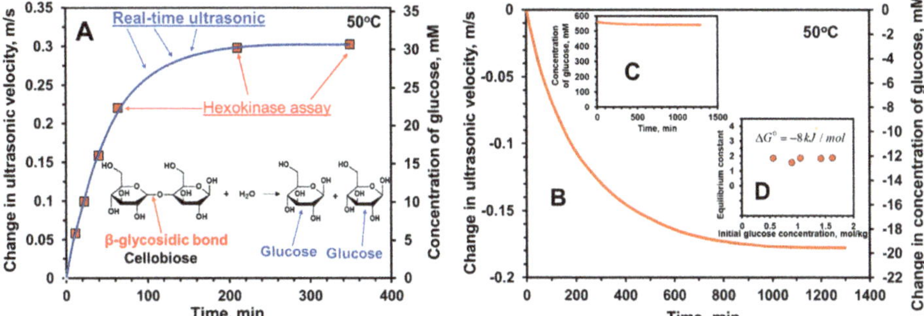

Figure 13. Real-time ultrasonic profile of hydrolysis and of reverse reaction of synthesis of cellobiose by enzyme β-glucosidase. Evaluation of reaction equilibrium constant and Gibbs energy.

Figure 13A. *Ultrasonic reaction profile of hydrolysis of cellobiose by β-glucosidase.*

Reaction (R8). 15.6 mmol/kg of cellobiose and 0.32 g/L of β-glucosidase. Squares represent the concentration of β-glycosidic bonds hydrolyzed (glucose formed) determined by colorimetric hexokinase assay by monitoring the production of NADPH measured at 340 nm (Megazyme Int. Ltd.) [8]. $\Delta a_r = 0.0126$ kg/mol was applied in calculations of the profile of concentration of glucose.

Figure 13B. *Time profile of concentration of glucose during reverse reaction (glucose condensation) catalyzed by β-glucosidase.*

20.3 g/L of β-glucosidase in aqueous solution of glucose at concentration 0.56 mol/kg, at 50 °C in 10 mM acetate buffer at pH 4.9. The change in concentration of glucose was calculated from the ultrasonic velocity profile [25].

Figure 13C. *Time profile of the overall concentration of glucose present in the reactional mixture during the reverse reaction catalyzed by β-glucosidase.*

Figure 13D. *Equilibrium constant, K_{eq}, of reaction of glucose condensation obtained at different concentrations of glucose.*

50 °C in 10 mM acetate buffer at pH 4.9. The presented ultrasonic velocity profiles were obtained at frequency 5.3 MHz with HR-US 102 ultrasonic spectrometer. No significant effect of frequency on the profiles were observed in the analyzed frequency range 2 to 20 MHz [25].

The concentrations of enzyme are given as mass (g) of Novozyme 188 liquid preparation (concentration of protein 184 mg/mL, density 1.23 g/mL at room temperature) per liter of solution.

Adapted with permission from Reference [25]. Copyright © 2011 Elsevier Inc.

8.1.2. Assessment of Enzyme Deactivation during Long-Time Reactions

Quantitative assessment of deactivation of enzyme during catalyzed processes is an important, however, often difficult and expensive task [78]. Figure 14 illustrates a direct ultrasonic assessment of stability of β-galactosidase in Cow & Gate First Infant milk at 20 °C during hydrolysis of lactose catalyzed by this enzyme over 24 h. The small concentration of enzyme utilized in the test was chosen to provide small (however measurable) change of concentration of the substrate (3% or 13 cm/s increase in ultrasonic velocity over 24 h), so that the hydrolysis occurred at nearly constant concentration of the substrate (as shown in the inset of Figure 14) [8]. The reaction was started by injection of a small (several μL) of diluted solution of enzyme (0.005 UU/g) to the measuring ultrasonic cell of HR-US 102 spectrometer containing 1.2 mL of infant milk, as described by Procedure 1. The concentration profile shown in Figure 14 was calculated from the measured ultrasonic velocity by applying Equations (5) and (7). The mainframe of Figure 14 also shows the reaction profile expected in an absence of enzyme deactivation (see [8] for details).

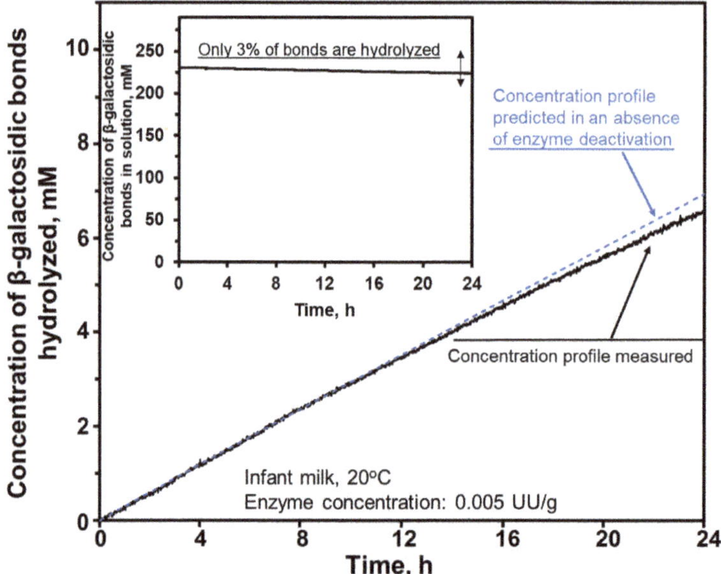

Figure 14. Ultrasonic assessment of deactivation of enzyme β-galactosidase during long time hydrolysis of lactose in milk.

Main frame. *24 h evolution of concentration of β-galactosidic bonds in Cow & Gate First Infant Milk during hydrolysis of the bonds by β-galactosidase.*

Reaction (R4). β-galactosidase from *Kluyveromyces lactis*, 0.005 UU/g, at 20 °C. The concentration (molarity) profile was calculated from the monitored change of ultrasonic velocity using Equations (7) and (10) and Δa_r = 0.0147 kg/mol. The black line represents the measured (HR-US 102 spectrometer) profile. The blue line represents the reaction profile expected in an absence of enzyme deactivation (see [8] for details).

Inset. *Total concentration of β-galactosidic bonds present in the reactional mixture during hydrolysis in Cow & Gate First Infant Milk.*

Calculated from the same as in the main frame ultrasonic profile according to Equation (5), where c^0 represent the concentration of β-galactosidic bonds of lactose and GOS in milk.

The ultrasonic measurements were collected in the frequency range 2 to 15 MHz. As no significant effects of frequency on the ultrasonic reaction profiles were observed, only the data at 5 MHz are presented. Adapted with permission from Reference [8]. Copyright © 2016 American Chemical Society.

According to the figure, at times above 16 h the reaction profile deviates from the one predicted in an absence of enzyme deactivation, which could be interpreted as a beginning of progressive enzyme deactivation.

8.2. Titrations in Analysis of Effects of Enzyme Concentration

The effect of concentration of enzyme on its activity is an important factor in the design of enzyme based formulations. It is commonly assessed by measuring the initial reaction rates in a set of samples containing enzyme of different concentrations. Figure 15 illustrates an alternative methodology

for the assessment of the effects of concentration of enzyme β-galactosidase on its activity in milk.
The methodology utilizes titration capabilities of HR-US technique. The measurements were performed
by titrating the infant milk in the ultrasonic cell with concentrated solution of β-galactosidase at 20 °C
using pre-programmed precision injections by HR-US 102 titration accessory. Each injection was
followed by a short period of stirring and a period without stirring. The mainframe of the figure
represents ultrasonic velocity profiles (secondary Y axis) during time intervals between the stirring
periods and subsequent injections. The circles on the inset in Figure 15 represent the initial rates
of hydrolysis, r^0, calculated from the slopes of linear parts of the dependences at each step of the
titration (see [8] for details). According to the figure, the initial reaction rate, r^0, increases linearly with
concentration of the enzyme, thus, indicating that the specific activity of the β-galactosidase in the
milk does not depend on its concentration within the analyzed concentration range.

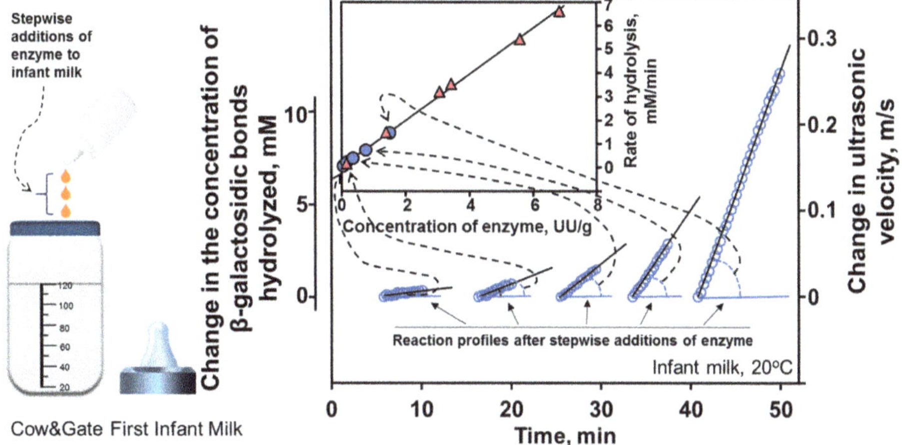

Figure 15. Ultrasonic assessment of effects of enzyme (β-galactosidase) concentration on its activity in
hydrolysis of lactose in milk.

Main frame.   *Time profiles of ultrasonic velocity in Cow & Gate First Infant Milk after automatic stepwise
additions of β-galactosidase (Kluyveromyces lactis) at 20 °C.*

The change in the concentration (molarity) of β-galactosidic bonds (lactose and GOS, Reaction
(R4)) hydrolyzed was calculated from the measured increase in ultrasonic velocity (HR-US102PT
spectrometer) according to Equations (10) and $\Delta a_r = 0.0147$ kg/mol.

Inset. *The initial rates of hydrolysis.*

The circles represent the initial rates of hydrolysis, obtained from the slopes shown in the mainframe
corrected for 'inhibition' effects [8]. Triangles represent the initial reaction rates obtained from separately
measured ultrasonic progress curves as those shown in Figure 6B.
The ultrasonic measurements were collected in the frequency range 2 to 15 MHz. As no significant
effects of frequency on the ultrasonic reaction profiles were observed, only the data at 5 MHz are
presented. Adapted with permission from Reference [8]. Copyright © 2016 American Chemical Society.

8.3. Temperature Profiling

Reversible and irreversible effects of temperature on activity of enzymes is one of the key
properties determining the enzymes applicability. They are commonly assessed by measuring the
initial reaction rates in a set of samples at different temperatures. Figure 16 illustrates an alternative
methodology, based on the measurements in a single sample utilizing the programmed temperature
profiling capabilities of HR-US technique for the enzyme β-galactosidase (Reaction (R4)) in milk [8].

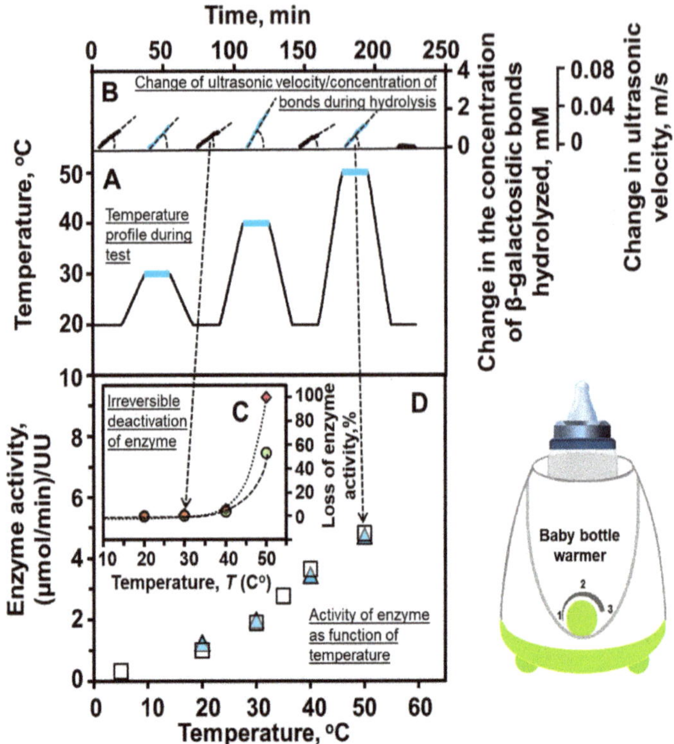

Figure 16. Ultrasonic assessment of reversible and non-reversible effects of temperature on activity of β-galactosidase in milk.

Effect of temperature on the rate of hydrolysis of β-galactosidic bonds (lactose and GOS, Reaction (R4)) in Cow & Gate First Infant Milk catalyzed by β-galactosidase (concentration 0.042 UU/g), reversible and irreversible deactivation of enzyme.

Figure 16A. *Temperature of the milk in ultrasonic cell in during single sample programmed temperature profiling.*

Figure 16B. *Ultrasonic velocity profiles at time intervals of constant temperature.*

HR-US-102PT ultrasonic spectrometer.

Figure 16C. *Irreversible loss of enzyme activity with temperature.*

Diamonds: The fractions of deactivated enzyme measured at subsequent (to the indicated temperatures) 20 °C time intervals, and presented as % of activity at the precluding 20 °C interval. Circles. The fractions of deactivated enzyme at each elevated temperature, calculated as the average of the fractions for the precluding and the subsequent 20 °C interval.

Figure 16D. *Effect of temperature on activity of active enzyme.*

Triangles: Calculated from the slopes ultrasonic velocity vs. time (B). Squares: the enzyme activities obtained from separately measured reaction profiles (Figure 2) [6]. Adapted with permission from Reference [8]. Copyright © 2016 American Chemical Society.

The measuring and the reference cells of HR-US 102 spectrometer were filled with 1.1 mL of infant milk. After equilibration at temperature 20 °C, 2 µL of β-galactosidase from Kluyveromyces lactis (aqueous solution, 21.6 UU/g, see figure legend for details) was added to the ultrasonic measuring cell, using Procedure 1. After the stirring of the sample using a mechanical mini stirrer for 30 s, the temperature profile (20 min time intervals at 20–30–20–40–20–50–20 °C, Figure 16A) was activated. The measured 'ultrasonic velocity vs. time profiles' (Figure 16B) provided the concentration of β-galactosidic bonds hydrolyzed, and the reaction rates r^0, calculated from the slopes ultrasonic velocity vs. time using Equations (8) and (10) with correction for inhibition effects (see [8] for details). The diamonds on Figure 16C illustrate the fraction of non-reversibly deactivated enzyme caused by heating from 20 °C to a particular temperature and subsequent cooling to 20 °C measured as a reduction of r^0 at 20 °C. The circles represent the fractions of deactivated enzyme at each elevated temperature, estimated as the average of the fractions for the precluding and the subsequent 20 °C interval. Figure 16D represents the level of activity of active enzyme at each elevated temperature interval calculated as a ratio of r^0 to the concentration of active enzyme expressed with ultrasonic activity units, UU (see Figure 6 legend for UU definition). The enzyme activities obtained from separately measured reaction profiles (e.g., Figure 6A) were added to the figure as opened squares. The highest specific activity was observed at 50 °C, which exceeds by approximately 2.5 times the activity at 30 °C. This agrees with previous results in buffers for chromogenic substrate ONPG [79]. Figure 16 shows that an exposure of the reaction mixture to temperature of 40 °C and above causes progressive irreversible deactivation of enzyme, which is particularly fast at 50 °C. The ultrasonic attenuation, measured during the test, has decreased with temperature, however, recovered to the original values at each 20 °C interval, thus, indicating an absence of irreversible effects of temperature on the milk microstructure during the test. The illustrated results allow optimization of the efficiency of commercial enzyme formulations for lactose-intolerant infants (see [8] for details).

9. Monitoring of Ultrasonic Attenuation

9.1. Structural Rearrangements and Particle Sizing

Figure 17 represents real-time ultrasonic monitoring of enzymatic removal of the 'hairy' layer of hydrated protein nanoparticles (casein micelles, 120 nm diameter approx.), followed by particle aggregation and gelation in milk at 30 °C [80]. The κ-casein 'hairy' layer provides steric stability of the particle dispersion. Its enzymatic removal leads to particle aggregation and formation of particle gel, which is utilized in cheese making (renneting process).

The measurements were performed according to Procedure 2 using HR-US 102 ultrasonic spectrometer in multifrequency regime in which the data measured at several pre-selected frequencies between 2 and 15 MHz were collected simultaneously. The enzyme (rennet, chymosin) was added to milk reconstituted from skim milk powder. The volume fraction of micelles, composed of casein and water, was 0.1. The ultrasonic attenuation profiles for different frequencies are presented in the inset and the profile for frequency 14.5 MHz is plotted in the main frame. This profile was recalculated into the evolution of the 'average' (over ultrasonic scattering profile, attenuation vs. size, see examples of the profiles in [30,41,81]) size of the protein particles during the process using the particle sizing module of HRUS 102 software, which utilized the equations as those discussed in Appendix A.3. The volume fraction of casein micelles was assumed to be constant. The physical parameters of the casein micelles and of the continuous medium utilized in these calculations were taken from data published by Griffin et al. [82]. Minor amendments of some parameters were introduced to account for small differences in temperature to which the parameters were originally attributed and the temperature at which the measurements were performed [80].

The particle size scale is presented on the right Y axis of the figure. According to the figure the average size of the particles decreases within first 8 min of reaction. The decrease in diameter from the initial size (120 nm [83]) is approximately 20 nm, which corresponds to the expected size of the particle 'hairy' layer [83]. This is followed by the increase of the size, which demonstrated the beginning of the aggregation of 'bald' particles. At the reaction time 30 min, further aggregation of particles results in a formation of gel network, as indicated by rheological (rise of shear storage modulus) data [80].

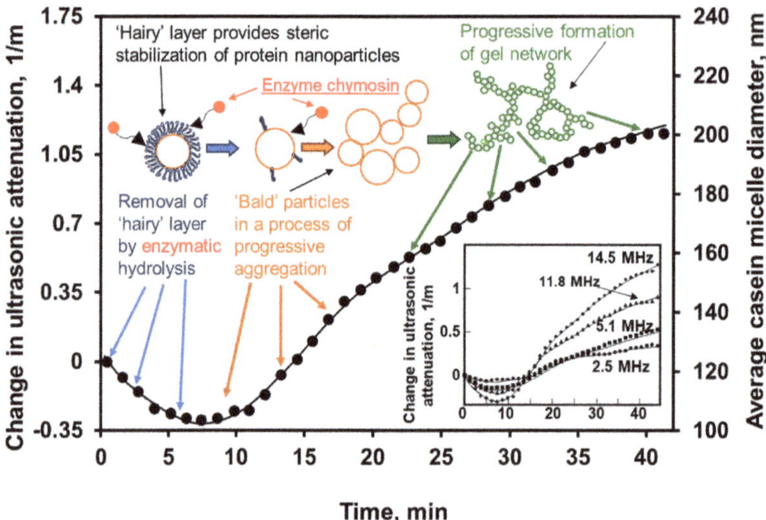

Figure 17. Ultrasonic real-time monitoring of evolution of size of protein nano particles during enzymatic 'haircut'.

Main frame. *Ultrasonic attenuation and particle size profile at 14.5 MHz in milk during enzymatic (chymosin) hydrolysis of the hairy layer of hydrated protein nanoparticles.*

Casein micelles, volume fraction 0.1 in milk at 30 °C. The average particle size was calculated using HR-US particle size software module.

Inset. *Effect of frequency on ultrasonic attenuation profiles of the hydrolysis.*

Measurements with HR-US 102 spectrometer (Ultrasonic Scientific, Dublin, Ireland). Adapted with permission from Reference [80]. Copyright © Proprietors of Journal of Dairy Research 2005.

9.2. Fast Chemical Kinetics

Figure 7C (inset) represents the attenuation profile of hydrolysis of peptide bonds of β-lactoglobulin (Reaction (R5)) catalyzed by protease α-chymotrypsin in 0.1 M phosphate buffer at pH 7.8) measured with HR-US 102 ultrasonic spectrometer using Procedure 1. As described in the Section 6.3 at pH 7.8 this reaction results in production of protonated, $-NH_3^+$, and deprotonated, $-NH_2$ terminal amino groups of oligopeptides. The ionization constant, pK_a, of these groups is 7.7 [84]. Consequently, these groups can participate in the reaction of proton transfer between them and the ions of the phosphate buffer: $-NH_3^+ + HPO_4^{-2} \rightleftharpoons -NH_2 + H_2PO_4^-$ [85–87]. Ionizable groups of amino acid side chains with pK_a values close to pH 7.8, which are exposed to contact with water as a result of hydrolysis, can also participate in the proton exchange with the ions of the phosphate buffer [85–87].

For common experimental conditions the relaxation frequency of the process of proton transfer is expected to be within low MHz range [85–89]. This agrees with our results presented in Figure 7C according to which at frequencies below 10 MHz (approximately) the amplitude of velocity rise during the process depends on the frequency, and a substantial increase of $\frac{\alpha}{f^2}$ during the reaction is observed. It is interesting to note that in contrary to the phosphate buffer, Tris, Bis-tris, Tes or Tricine and similar buffers do not show similar excessive attenuation in solutions of proteins at neutral pH in the low MHz frequency range [85]. When Equations (A11) and (A12) are considered, the observed frequency dependence of the rise of velocity and of $\frac{\alpha}{f^2}$ during the reaction indicates that the relaxation frequency f_{rel} of the proton transfer is at the lower side of the analyzed frequency range (2.7 MHz to 15.5 MHz), or below it. At frequencies above 10 MHz (approximately), the frequency dependence of the rise of ultrasonic velocity and the amplitude of the rise of $\frac{\alpha}{f^2}$ vanish. At these high (relative to f_{rel}) frequencies, the relaxation process is 'frozen' as the period of oscillation of pressure and temperature in ultrasonic wave is significantly shorter than the relaxation time of the proton transfer. Although quantitative analysis of the evolution of frequency dependence of ultrasonic velocity and $\frac{\alpha}{f^2}$ during hydrolysis of β-lactoglobulin is outside of the scope of this paper, it shall be stated that such analysis provides new insights into the mechanisms of protein hydrolysis and can be used as a tool for real time monitoring of this reaction in phosphate buffer as well as real-time measurements of production of terminal amino groups of oligopeptides during hydrolysis of proteins.

Figure 5B illustrates the attenuation profile during the hydrolysis of maltodextrin. The decrease in attenuation cannot be explained only by the decrease in classical contribution to attenuation if 'low' frequency viscosities, η, of solutions maltodextrins maltodextrins [90–93] are applied to Equation (16). A relaxation process associated with conformational rearrangements within maltodextrins (see discussion of ultrasonic attenuation in aqueous solutions of carbohydrates in [94]) could be a reason for the observed attenuation decrease. Detailed analysis of this phenomenon shall provide additional capabilities for ultrasonic monitoring and analysis of hydrolysis of oligo and polysaccharides.

10. Conclusions

High-resolution ultrasonic spectroscopy can be successfully employed for real-time, non-destructive analysis of reactions catalyzed by enzymes and other catalysts in solutions and complex liquid dispersions. This technique provides precision reaction progress curves (concentration of reactant/product vs. time) over the whole course of reaction, and also can be utilized for analysis of structural rearrangements (change of particle size, aggregation) during reactions. It does not require optical transparency or optical markers and is applicable for reactions in continuous media and in micro/nano bioreactors (e.g., nanodroplets of microemulsions). The accuracy of commercial high-resolution ultrasonic spectrometers in measurements of change of ultrasonic velocity (down to 0.2 mm/s [23]) corresponds to μM level of precision in monitoring of the evolution of the concentrations of reactants and products [5,8,25]. This precision stands for mixtures with low and high concentration of substrates, which allows efficient assessment of stability of enzymes/catalysts and analysis of 'low yielding' reactions. The high precision can be used for measurements of the detailed 'reaction rate vs. concentration' profiles over the course of reaction, utilized in advanced modelling of reaction kinetics and inhibition effects. Other useful capabilities of the technique include measurements with programmable temperature profiles for assessments of reversible and non-reversible effects of thermal history on enzyme activity in a single sample. As ultrasonic measurements characterize the properties of the bulk medium, the unwanted effects of surfaces, often associated with reflectance spectroscopies and electrode techniques, are excluded. Since most of the chemical reactions in liquids are accompanied by hydration or solvation effects the described methodology should be applicable to a variety of reactions catalyzed by enzymes and other catalysts in various media. Importantly, the ultrasonic analysis can be carried out directly on intact samples with native substrates, thus allowing optimal application of catalysts in targeted media. Overall, the capabilities of this technique could make it a valuable tool in the field of catalysis based technologies.

Acknowledgments: We are grateful to Evgeny Koudryashov for his contribution with ultrasonic measurements of reactions of synthesis of ATP and hydrolysis of maltodextines, Shailesh Kumar for his contribution with ultrasonic measurements in microemulsions, Mark Dizon for his technical assistance, Rian Lynch for his assistance in editing the text, and to Sonas Technologies Ltd. for providing HR-US ultrasonic spectrometers, accessories and technical support. This research has been supported by 06RDTMFRC444 and 13F454 Grants from the Department of Agriculture, Food and the Marine of Ireland, MASF315 Grant from Science Foundation Ireland, H2020 EU Grant 690898 FORMILK, Postgraduate fellowship for Margarida Altas from School of Chemistry of University College Dublin.

Author Contributions: V.B. designed the structure and wrote the paper, M.C.A. contributed to data acquisition and data analysis, drafted the manuscript sections and prepared the figures. Both authors contributed to editing of the manuscript.

Conflicts of Interest: The authors declare no conflict of interest.

Appendix A.

Appendix A.1. Degree of Polymerization and Average Molar Mass

For a mixture of polymers consisting of monomers linked with the same covalent bond the average degree of polymerization, $\overline{DP}$, and the average molar mass, $\overline{M}$ are defined as:

$$\overline{DP} = \frac{\sum\limits_{n=1}^{k} c_n n}{\sum\limits_{n=1}^{k} c_n}; \quad \overline{M} = \frac{\sum\limits_{n=1}^{k} c_n M_n}{\sum\limits_{n=1}^{k} c_n} \tag{A1}$$

where c_n is the concentration (number of moles in 1 kg of mixture) of molecules consisting of n monomeric units, M_n is their molar mass, and k is the maximum number of monomeric units in the polymer molecules. For linear polymers, the total concentration (number of moles in 1 kg of mixture) of bonds, $c_b(t)$, between the monomers at time t is given as:

$$c_b(t) = \sum\limits_{n=1}^{k} c_n(n-1) = \sum\limits_{n=1}^{k} c_n n - \sum\limits_{n=1}^{k} c_n \tag{A2}$$

The total number of monomer units in all polymer molecules in 1 kg of the mixture, c_m, is given as:

$$c_m = \sum\limits_{n=1}^{k} c_n n \tag{A3}$$

The parameter c_m is not affected by the breaking of the bonds between monomeric units within the molecules of polymer and, therefore, is constant over time. Following the above Equation (A2) can be rearranged as:

$$\sum\limits_{n=1}^{k} c_n = c_m - c_b(t) \tag{A4}$$

The concentration $c_b(t)$ can be presented as:

$$c_b(t) = c_b{}^0 - c_{bh}(t) \tag{A5}$$

where $c_b{}^0$ is the concentration of the bonds at time zero ($t = 0$) and $c_{bh}(t)$ is the concentration of bonds broken during the time interval between time zero and time t. At $t = 0$ $c_{bh}(t) = 0$. Application of Equations (A3)–(A5) to the first of Equation (A1) results in:

$$\overline{DP} = \frac{\overline{DP}^0}{1 + c_{bh}(t)\frac{\overline{DP}^0 - 1}{c_b{}^0}} \tag{A6}$$

where $\overline{DP}^0$ is the average degree of polymerization at time zero ($t = 0$).

The sum $\sum_{n=1}^{k} c_n M_n$ in the second Equation (A1) represents the total mass of the molecules of the polymer in 1 kg of the mixture, which is the polymer weight fraction in the mixture, w_p:

$$\sum_{n=1}^{k} c_n M_n = w_p \tag{A7}$$

This and Equations (A1)–(A5) provide the following useful relationship:

$$\frac{\overline{DP}^0 - 1}{c_b{}^0} = \frac{\overline{M}^0}{w_p{}^0} \tag{A8}$$

where $w_p{}^0$ and $\overline{M}^0$ are the polymer weight fraction and the average molar mass at time zero ($t = 0$).

Derivation of an equivalent to Equation (A6) relationship for average molar mass, $\overline{M}$, requires an appropriate relationship between the weight fraction of polymer, w_p, and the concentration of bonds broken by the reaction, $c_{bh}(t)$. For reactions of hydrolysis, which are accompanied by a consumption of one mole of water molecule per one mole of bonds hydrolyzed, $w_p = w_p{}^0 + M_{H_2O} c_{bh}(t)$, where M_{H_2O} is the molar mass of water. Application of this relationship to Equations (A1)–(A5) and (A7) provides:

$$\overline{M} = \frac{\overline{M}^0 - M_{H_2O}}{1 + c_{bh}(t)\frac{\overline{M}^0}{w_p{}^0}} + M_{H_2O} \tag{A9}$$

For hydrolysis of polymers composed of the monomer units of the same molar mass the following relationship is correct $M_n = (M_1 - M_{H_2O})n + M_{H_2O}$, where M_1 is the molar mass of the free monomer. Application of this relationship to Equation (A1) results in:

$$\begin{aligned} \overline{M} &= (M_1 - M_{H_2O})\overline{DP} + M_{H_2O} \\ \overline{M}^0 &= (M_1 - M_{H_2O})\overline{DP}^0 + M_{H_2O} \end{aligned} \tag{A10}$$

These equations allow easy recalculations of the average degree of polymerization of such polymers to their average molar mass, which is especially useful for calculations of $\overline{DP}^0$ or $\overline{M}^0$.

Appendix A.2. Relaxation Contribution to Ultrasonic Velocity and Attenuation

For a relaxation process described by Reaction (R1) and characterized by the relaxation time τ, the contribution to attenuation caused by this process in dilute solutions is given by the following relationship [17,18]:

$$\frac{\alpha_{rel}}{f^2} = \frac{\pi \rho u \, \Delta V_S{}^2 \Gamma}{RT} \frac{f_{rel}}{f_{rel}^2 + f^2} \tag{A11}$$

where R is the gas constant, $\Gamma = \left(\sum_{i=1}^{k} \frac{a_i{}^2}{c_{A_i}} + \sum_{i=1}^{l} \frac{b_i{}^2}{c_{B_i}}\right)^{-1}$ is the term representing the equilibrium concentrations of the reactants, c_{A_i}, and products, c_{B_i}, and the stoichiometric coefficients a_i and b_i of Reaction (R1), $f_{rel} = \frac{1}{2\pi\tau}$ is the relaxation frequency of the process, determined by the kinetic constants of the reaction and the concentrations of the reactants and products, ΔV_S is the volume effect of the reaction at adiabatic conditions given by: $\Delta V_S = \Delta V_T - \frac{e}{c_p}\Delta H$, where ΔV_T is the volume effect of the reaction (change of volume of the solution as a result of transferring of $a_1, a_2, \ldots, a_k$ moles of reactants $A_1, A_2, \ldots, A_k$ to $b_1, b_2, \ldots, b_k$ moles of products $B_1, B_2, \ldots, B_k$) at constant temperature and

ΔH is the enthalpy of reaction. The contribution of the relaxation process to ultrasonic velocity, u_{rel}, at $|u_{rel}| \ll u$ (e.g., dilute solution) is given as:

$$u_{rel} = -\frac{\rho u^3}{2RT}\Delta V_S^2 \Gamma \frac{f_{rel}^2}{f_{rel}^2 + f^2} = -\frac{f_{rel}u^2}{2\pi}\frac{\alpha_{rel}}{f^2} \tag{A12}$$

According to Equations (A11) and (A12) the relaxation contribution $\frac{\alpha_{rel}}{f^2}$ is positive and the contribution u_{rel} is negative. The dependencies of both contributions on the frequency have an S-shape profile with plateaus at low ($f \ll f_{rel}$) and high ($f \gg f_{rel}$) frequencies and half transition point at $f = f_{rel}$. At measuring frequencies significantly lower than f_{rel} the relaxation contribution to the ultrasonic attenuation is given as $\frac{\alpha_{rel}}{f^2} = \frac{\pi \rho u}{RT}\frac{\Delta V_S^2 \Gamma}{f_{rel}}$ and to the ultrasonic velocity as $u_{rel} = -\frac{\rho u^3}{2RT}\Delta V_S^2 \Gamma$. At frequencies close to f_{rel} the absolute value of both contributions decreases with frequency and vanish at frequencies far above f_{rel}. At these high frequencies, the relaxation process is 'frozen' as its relaxation time is much longer than the period of oscillations of pressure and temperature in the ultrasonic wave.

Appendix A.3. Ultrasonic Particle Sizing

Heterogeneous dispersions produce 'scattering' contribution to ultrasonic velocity and attenuation, which is a function of the particle's size and the volume fraction [95–98]. Generally, analysis of the scattering from particles requires numerical solutions. Nevertheless, in the long wavelength regime, (i.e., when the wavelength of ultrasound is much longer than the particle radius), explicit expressions for the ultrasonic scattering in dispersions could be utilized [99]. The basic mechanism of the interaction of ultrasonic wave with particles in dispersions in this regime can be illustrated with two major scattering contributions, thermoelastic and viscoinertial [22,99]. These contributions result from the 'scattering' of the incident ultrasonic waves into the thermal and viscous waves on the border between the particle and the continuous medium. The thermoelastic and viscoinertial mechanisms normally dominate the scattering contribution to ultrasonic attenuation in dispersions of micron and submicron size range and for the operational frequency range of high-resolution ultrasonic spectrometers (1 to 20 MHz). For the long wavelength regime, the approximation formulas describing the ultrasonic velocity and attenuation in dispersions of spherical particles can be presented as [100,101]:

$$K^2 = k_1^2\left(1 - \frac{3\phi i B_0}{k_1^3 r^3}\right)\left(1 - \frac{9\phi i B_1}{k_1^3 r^3}\right) \tag{A13}$$

where:

$K = \frac{\omega}{u} + i\alpha$ is a complex propagation constant of the dispersion,

u is the ultrasonic velocity and α is the ultrasonic attenuation coefficient in the dispersion,

ϕ is the volume fraction of the dispersed phase,

$i = \sqrt{-1}$,

r is the particle radius,

$\omega = 2\pi f$ is the angular frequency, f is the ultrasonic frequency,

$k_1 = \frac{\omega}{u_1} + i\alpha_1$ is the complex propagation constant of the continuous phase,

u_1 is the ultrasonic velocity and α_1 is the attenuation in the continuous phase,

$$B_0 = \frac{ik_1^3 r^3}{3}\left(\frac{\rho_1 k_2^2}{\rho_2 k_1^2} - 1\right) - \frac{ik_1^3 r^3 \xi H}{z_1^2}\left(1 - \frac{\beta_2 \rho_1 c_{p1}}{\beta_1 \rho_2 c_{p2}}\right)^2,$$

$$B_1 = \frac{-ik_1^3 r^3}{9}\left(\frac{(\rho_1 - \rho_2)(1 + T_v + is)}{(\rho_2 + \rho_1 T_v + i\rho_1 s)}\right),$$

$$\frac{1}{H} = \frac{1}{(1 - iz_1)} - \frac{\kappa_1}{\kappa_2}\frac{\tan(z_2)}{\tan(z_2) - z_2},$$

$$s = \frac{9\delta}{4r}\left(1 + \frac{\delta}{r}\right),$$

$$T_v = \frac{1}{2} + \frac{9\delta}{4r},$$

$$z_1 = (1+i)\frac{r}{\delta_1}, z_2 = (1+i)\frac{r}{\delta_2},$$

$$\zeta = \frac{T\beta_1{}^2 u_1{}^2}{c_{p1}},$$

$$\delta = \sqrt{\frac{2\eta_1}{\omega\rho_1}}, \ \delta_1 = \sqrt{\frac{2\kappa_1}{\omega c_{p1}\rho_1}}, \ \delta_2 = \sqrt{\frac{2\kappa_2}{\omega c_{p2}\rho_2}},$$

$k_2 = \frac{\omega}{u_2} + i\alpha_2$ is the complex propagation constant of the dispersed phase,

u_2 is the ultrasonic velocity and α_2 is the attenuation coefficient in the dispersed phase,

η_1 is the viscosity of the continuous phase,

$\beta_1 = \rho_1 e_1$, $\beta_2 = \rho_2 e_2$ are the coefficients of volume expansion for the continuous phase and the dispersed phase,

c_{p1}, c_{p2} are the specific heat capacities of the continuous and dispersed phase at constant pressure,

ρ_1, ρ_2 are the densities of the continuous and dispersed phases,

κ_1, κ_2 are thermal conductivities of the continuous and dispersed phases,

T is temperature in K.

Appendix A.4. Materials and Instruments Utilized in Previously Unpublished Illustrations

All aqueous solutions were prepared by weight using purified water (Milli-Q gradient ultrapure purification system, conductivity 0.056 µS/cm, resistivity 18 MΩ cm, TOC max. value 100 µg/L). Phosphate buffers were prepared from dibasic (anhydrous BioUltra cat. 60353) and monobasic (anhydrous BioUltra cat. 60218) potassium phosphate.

Ultrasonic measurements were performed with HR-US 102 range spectrometers and their accessories from Sonas Technologies Ltd. (Dublin, Ireland).

Figures 3 and 7.

β-lactoglobulin from bovine milk from Sigma-Aldrich Co. (St. Louis, MO, USA), cat. L3908, lyophilized powder. α-chymotrypsin from bovine pancreas, Sigma-Aldrich Co, cat. C3142, Type VII, lyophilized powder, with a declared activity of ≥64 U/mg protein. 2,4,6-Trinitrobenzenesulfonic acid (TNBS) from Sigma-Aldrich Co., cat. P2297.

Figures 3 and 8.

ADP, adenosine 5′-diphosphate sodium salt, from Sigma-Aldrich Co., cat. A2754. Phosphocreatine disodium salt hydrate from Sigma-Aldrich Co., cat. P7936. Creatine phosphokinase from rabbit muscle, Sigma-Aldrich Co., cat. C3755, type I, lyophilized powder, with a declared activity of ≥150 U/mg protein. Gly-gly (diglycine) from Sigma-Aldrich Co., cat. G1002, BSA from Sigma-Aldrich Co., cat. B4287. Magnesium acetate Sigma-Aldrich Co., cat. 22864-8.

Figures 5 and 11.

Maltodextrin from Sigma-Aldrich Co., cat. 419680, 14.5 dextrose equivalents (corresponds to average degree of polymerization 7.55). α-amylase from *Bacillus* sp., Sigma-Aldrich Co., cat. A6380, Type II-A, lyophilized powder, with a declared activity of 2680 U/mg protein, concentration 0.016 g/L. BCA assay: Bicinchoninic acid (4,4′-dicarboxy-2,2′-biquinoline) from Sigma-Aldrich Co., cat. 391778; Na_2CO_3 anhydrous powder from Sigma-Aldrich Co., cat. 451614; aspartic acid from Sigma-Aldrich Co., cat. A93100; $CuSO_4$ anhydrous powder from Sigma-Aldrich Co. cat. 451657.

References

1. Hall, A.M.; Chouler, J.C.; Codina, A.; Gierth, P.T.; Lowe, J.P.; Hintermair, U. Practical aspects of real-time reaction monitoring using multi-nuclear high resolution FlowNMR spectroscopy. *Catal. Sci. Technol.* **2016**, *6*, 8406–8417. [CrossRef]
2. Kudryashov, E.; Smyth, C.; O'Driscoll, B.; Buckin, V. High-Resolution Ultrasonic Spectroscopy for analysis of chemical reactions in real time. *Spectroscopy* **2003**, *18*, 26–32.

3. Danieli, E.; Perlo, J.; Duchateau, A.; Verzijl, G.; Litvinov, V.; Blümich, B.; Casanova, F. On-Line Monitoring of Chemical Reactions by using Bench-Top Nuclear Magnetic Resonance Spectroscopy. *ChemPhysChem* **2014**, *15*, 3060–3066. [CrossRef] [PubMed]

4. Bernstein, M.A.; Štefinović, M.; Sleigh, C.J. Optimising reaction performance in the pharmaceutical industry by monitoring with NMR. *Magn. Reson. Chem.* **2007**, *45*, 564–571. [CrossRef] [PubMed]

5. Li, Y.; Buckin, V. Ultrasonic real-time monitoring of the process of decomposition of hydrogen peroxide in aqueous solutions. *Anal. Methods* **2016**, *8*, 4828–4834. [CrossRef]

6. Baltes, W. *Rapid Methods for Analysis of Food and Food Raw Material*; Behr's Verlag: Hamburg, Germany, 1990.

7. Wiss, J.; Länzlinger, M.; Wermuth, M. Safety improvement of a Grignard reaction using on-line NIR monitoring. *Org. Process Res. Dev.* **2005**, *9*, 365–371. [CrossRef]

8. Caras Altas, M.; Kudryashov, E.; Buckin, V. Ultrasonic monitoring of enzyme catalysis. Enzyme activity in formulations for lactose intolerant infants. *Anal. Chem.* **2016**, *88*, 4714–4723. [CrossRef] [PubMed]

9. Wahler, D.; Reymond, J.-L. Novel methods for biocatalyst screening. *Curr. Opin. Chem. Biol.* **2001**, *5*, 152–158. [CrossRef]

10. Bothner, B.; Chavez, R.; Wei, J.; Strupp, C.; Phung, Q.; Schneemann, A.; Siuzdak, G. Monitoring enzyme catalysis with mass spectrometry. *J. Biol. Chem.* **2000**, *275*, 13455–13459. [CrossRef] [PubMed]

11. Mello, L.D.; Kubota, L.T. Review of the use of biosensors as analytical tools in the food and drink industries. *Food Chem.* **2002**, *77*, 237–256. [CrossRef]

12. Povey, M.J.; Mason, T.J. *Ultrasound in Food Processing*; Springer: Berlin, Germany, 1998.

13. Buckin, V. Application of High-Resolution Ultrasonic Spectroscopy for analysis of complex formulations. Compressibility of solutes and solute particles in liquid mixtures. *IOP Conf. Ser. Mater. Sci. Eng.* **2012**, *42*, 1–18. [CrossRef]

14. Holmes, M.; Povey, M. Ultrasonic Particle Sizing in Emulsions. *Ultrasound Food Proc. Recent Adv.* **2017**, 27–64. [CrossRef]

15. Kaatze, U.; Eggers, F.; Lautscham, K. Ultrasonic velocity measurements in liquids with high resolution—Techniques, selected applications and perspectives. *Meas. Sci. Technol.* **2008**, *19*, 1–21. [CrossRef]

16. Buckin, V.; Kudryashov, E.; O'Driscoll, B. An alternative spectroscopy technique for biopharmaceutical applications. *Pharm. Technol. Eur.* **2002**, *14*, 33–37.

17. Eigen, M.; De Maeyer, L. *Techniques of Organic Chemistry*; Friess, S.L., Lewis, E.S., Weisberger, A., Eds.; Investigation of Rates and Mechanisms of Reactions; Relaxation Methods Interscience: New York, NY, USA, 1963; Volume 8.

18. Stuehr, J.; Yeager, E. *Physical Acoustics*; Mason, W.P., Ed.; Academic Press: New York, NY, USA, 1965; Volume 2.

19. Smyth, C.; Kudryashov, E.; O'Driscoll, B.; Buckin, V. High-resolution ultrasonic spectroscopy for analysis of industrial emulsions and suspensions. *JALA J. Assoc. Lab. Autom.* **2004**, *9*, 87–90. [CrossRef]

20. Holmes, M.; Southworth, T.; Watson, N.; Povey, M. *Enzyme Activity Determination Using Ultrasound*; IOP Publishing: Bristol, UK, 2014; p. 012003.

21. Resa, P.; Elvira, L.; De Espinosa, F.M.; Gómez-Ullate, Y. Ultrasonic velocity in water–ethanol–sucrose mixtures during alcoholic fermentation. *Ultrasonics* **2005**, *43*, 247–252. [CrossRef] [PubMed]

22. Povey, M.J. *Ultrasonic Techniques for Fluids Characterization*; Academic Press Limited: London, UK, 1997.

23. Buckin, V.; O'Driscoll, B. Ultrasonic waves and material analysis: Recent advances and future trends. *LabPlus Int.* **2002**, *16*, 17–21.

24. Buckin, V.; Smyth, C. High-resolution ultrasonic resonator measurements for analysis of liquids. *Semin. Food Anal.* **1999**, *4*, 113–130.

25. Resa, P.; Buckin, V. Ultrasonic analysis of kinetic mechanism of hydrolysis of cellobiose by beta-glucosidase. *Anal. Biochem.* **2011**, *415*, 1–11. [CrossRef] [PubMed]

26. Niemeyer, K. Ultrasonic Methods for Analytical Determination of Pancreatic Enzyme Activities in Pharmaceutical Preparations. Ph.D. Thesis, PHD Thesis Fakultät für Lebenswissenschaften der Technischen Universität Carolo-Wilhelmina, Braunschweig, Germany, 2012.

27. Jager, M.; Kaatze, U.; Kudryashov, E.; O'Driscoll, B.; Buckin, V. New capabilities of high-resolution ultrasonic spectroscopy: Titration analysis. *Sperctoscopy* **2005**, *20*, 20–26.

28. Van Durme, K.; Delellio, L.; Kudryashov, E.; Buckin, V.; Van Mele, B. Exploration of high-resolution ultrasonic spectroscopy as an analytical tool to study demixing and remixing in poly(N-isopropyl acrylamide)/water solutions. *J. Polym. Sci. Part B Polym. Phys.* **2005**, *43*, 1283–1295. [CrossRef]

29. Hickey, S.; Lawrence, M.J.; Hagan, S.A.; Buckin, V. Analysis of the phase diagram and microstructural transitions in phospholipid microemulsion systems using high-resolution ultrasonic spectroscopy. *Langmuir* **2006**, *22*, 5575–5583. [CrossRef] [PubMed]

30. Hickey, S.; Hagan, S.A.; Kudryashov, E.; Buckin, V. Analysis of phase diagram and microstructural transitions in an ethyl oleate/water/Tween 80/Span 20 microemulsion system using high-resolution ultrasonic spectroscopy. *Int. J. Pharm.* **2010**, *388*, 213–222. [CrossRef] [PubMed]

31. Buckin, V.; Hallone, S.K. Ultrasonic Characterisation of W/O Microemulsions—Structure, Phase Diagrams, State of Water in Nano–Droplets, Encapsulated Proteins, Enzymes. In *Microemulsions—An Introduction to Properties and Applications*; InTech: London, UK, 2012; pp. 33–66.

32. Bainor, A.; Chang, L.; McQuade, T.J.; Webb, B.; Gestwicki, J.E. Bicinchoninic acid (BCA) assay in low volume. *Anal. Biochem.* **2011**, *410*, 310–312. [CrossRef] [PubMed]

33. Johnston, D.; Shoemaker, S.; Smith, G.; Whitaker, J. Kinetic measurements of cellulase activity on insoluble substrates using disodium 2, 2′ bicinchoninate. *J. Food Biochem.* **1998**, *22*, 301–319. [CrossRef]

34. Adler-Nissen, J. Determination of the Degree of Hydrolysis of Food Protein Hydrolysates by Trinitrobenzenesulfonic Acid. *J. Agric. Food Chem.* **1979**, *27*, 1256–1262. [CrossRef] [PubMed]

35. Kharakoz, D.P. Partial Volumes and Compressibilities of Extended Polypeptide Chains in Aqueous Solution: Additivity Scheme and Implication of Protein Unfolding at Normal and High Pressure. *Biochemistry* **1997**, *36*, 10276–10285. [CrossRef] [PubMed]

36. Vorob'ev, M.; Levicheva, I.Y.; Belikov, V. Kinetics of the initial stage of milk protein hydrolysis by chymotrypsin. *Appl. Biochem. Microbiol.* **1996**, *32*, 219–222.

37. Goncharova, I. Computer simulation of proteolysis. Peptic hydrolysis of partially demasked β-Lactoglobulin. *Mol. Nutr. Food Res.* **1998**, *42*, 61–67.

38. Nihei, T.; Noda, L.; Morales, M.F. Kinetic Properties and Equilibrium Constant of the Adenosine Triphosphate-Creatine Transphosphorylase-catalyzed Reaction. *J. Biol. Chem.* **1961**, *236*, 3203–3209. [PubMed]

39. Pierce, A.D. *Acoustics: An Introduction to Its Physical Principles and Applications*; American Institute of Physics: New York, NY, USA, 1989.

40. Litovitz, T.A.; Davis, C.M. *Physical Acoustics*; Mason, W.P., Ed.; Academic Press: New York, NY, USA, 1964; Volume 2.

41. Ochenduszko, A.; Buckin, V. Real-time monitoring of heat-induced aggregation of β-lactoglobulin in aqueous solutions using high-resolution ultrasonic spectroscopy. *Int. J. Thermophys.* **2010**, *31*, 113–130. [CrossRef]

42. Hoshino, E.; Tanaka, A.; Kanda, T. Effects of a nonionic surfactant on the behavior of Bacillus amyloliquefaciens α-amylase in the hydrolysis of malto-oligosaccharide. *J. Surfactants Deterg.* **2006**, *9*, 63–68. [CrossRef]

43. Robyt, J.; French, D. Action pattern and specificity of an amylase from Bacillus subtilis. *Arch. Biochem. Biophys.* **1963**, *100*, 451–467. [CrossRef]

44. Yook, C.; Robyt, J.F. Reactions of alpha amylases with starch granules in aqueous suspension giving products in solution and in a minimum amount of water giving products inside the granule. *Carbohydr. Res.* **2002**, *337*, 1113–1117. [CrossRef]

45. Rodriguez-Colinas, B.; Fernandez-Arrojo, L.; Ballesteros, A.; Plou, F. Galactooligosaccharides formation during enzymatic hydrolysis of lactose: Towards a prebiotic-enriched milk. *Food Chem.* **2014**, *145*, 388–394. [CrossRef] [PubMed]

46. Jurado, E.; Camacho, F.; Luzon, G.; Vicaria, J.M. Kinetic models of activity for beta-galactosidases: Influence of pH, ionic concentration and temperature. *Enzyme Microb. Technol.* **2004**, *34*, 33–40. [CrossRef]

47. Martínez-Villaluenga, C.; Cardelle-Cobas, A.; Corzo, N.; Olano, A.; Villamiel, M. Optimization of conditions for galactooligosaccharide synthesis during lactose hydrolysis by β-galactosidase from Kluyveromyces lactis (Lactozym 3000 L HP G). *Food Chem.* **2008**, *107*, 258–264. [CrossRef]

48. Roefs, P.; de Kruif, K.G. Association behavior of native ß-lactoglobulin. *Biopolymers* **1999**, *49*, 11–20.

49. Berg, J.M.; Tymoczko, T.J.; Stryer, L. Catalytic Strategies. In *Biochemistry*; WH Freeman and Company: New York, NY, USA, 2007; pp. 241–274.

50. Lisak, K.; Toro-Sierra, J.; Kulozik, U.; Božanić, R.; Cheison, S.C. Chymotrypsin selectively digests β-lactoglobulin in whey protein isolate away from enzyme optimal conditions: Potential for native α-lactalbumin purification. *J. Dairy Res.* **2013**, *80*, 14–20. [CrossRef] [PubMed]

51. Maria, C.; Galvão, A.; Silva, A.F.S.; Custódio, M.F.; Monti, R.; De, R.; Giordano, C. Controlled hydrolysis of cheese whey proteins using trypsin and [alpha]-chymotrypsin. *Appl. Biochem. Biotechnol.* **2001**, *91*, 761–776.

52. Brot, F.E.; Bender, M.L. Use of the specificity constant of. alpha.-chymotrypsin. *J. Am. Chem. Soc.* **1969**, *91*, 7187–7191. [CrossRef]

53. Chae, H.J.; In, M.-J.; Kim, M.-H. Process development for the enzymatic hydrolysis of food protein: Effects of pre-treatment and post-treatments on degree of hydrolysis and other product characteristics. *Biotechnol. Bioprocess Eng.* **1998**, *3*, 35–39. [CrossRef]

54. Clemente, A. Enzymatic protein hydrolysates in human nutrition. *Trends Food Sci. Technol.* **2000**, *11*, 254–262. [CrossRef]

55. Korhonen, H. Milk-derived bioactive peptides: From science to applications. *J. Funct. Foods* **2009**, *1*, 177–187. [CrossRef]

56. Dawson, D.M. Creatine kinase from brain: Kinetic aspects. *J. Neurochem.* **1970**, *17*, 65–74. [CrossRef] [PubMed]

57. Tanzer, M.L.; Gilvarg, C. Creatine and Creatine Kinase Measurement. *J. Biol. Chem.* **1959**, *234*, 3201–3204. [PubMed]

58. Lin, S.-S.; Gurol, M.D. Catalytic Decomposition of Hydrogen Peroxide on Iron Oxide: Kinetics, Mechanism, and Implications. *Environ. Sci. Technol.* **1998**, *32*, 1417–1423. [CrossRef]

59. Watts, R.J.; Foget, M.K.; Kong, S.-H.; Teel, A.L. Hydrogen peroxide decomposition in model subsurface systems. *J. Hazard. Mater.* **1999**, *69*, 229–243. [CrossRef]

60. Petri, B.G.; Watts, R.J.; Teel, A.L.; Huling, S.G.; Brown, R.A. Fundamentals of ISCO Using Hydrogen Peroxide. In *In Situ Chemical Oxidation for Groundwater Remediation*; Siegrist, R.L., Crimi, M., Simpkin, T.J., Eds.; Springer: New York, NY, USA, 2011; pp. 33–88.

61. Zepp, R.G.; Faust, B.C.; Hoigne, J. Hydroxyl radical formation in aqueous reactions (pH 3–8) of iron(II) with hydrogen peroxide: The photo-Fenton reaction. *Environ. Sci. Technol.* **1992**, *26*, 313–319. [CrossRef]

62. Kwan, W.P.; Voelker, B.M. Decomposition of Hydrogen Peroxide and Organic Compounds in the Presence of Dissolved Iron and Ferrihydrite. *Environ. Sci. Technol.* **2002**, *36*, 1467–1476. [CrossRef] [PubMed]

63. Chu, L.; Wang, J.; Dong, J.; Liu, H.; Sun, X. Treatment of coking wastewater by an advanced Fenton oxidation process using iron powder and hydrogen peroxide. *Chemosphere* **2012**, *86*, 409–414. [CrossRef] [PubMed]

64. Chakinala, A.G.; Bremner, D.H.; Gogate, P.R.; Namkung, K.-C.; Burgess, A.E. Multivariate analysis of phenol mineralisation by combined hydrodynamic cavitation and heterogeneous advanced Fenton processing. *Appl. Catal. B Environ.* **2008**, *78*, 11–18. [CrossRef]

65. Doong, R.-A.; Chang, W.-H. Photodegradation of parathion in aqueous titanium dioxide and zero valent iron solutions in the presence of hydrogen peroxide. *J. Photochem. Photobiol. A Chem.* **1998**, *116*, 221–228. [CrossRef]

66. Bremner, D.H.; Burgess, A.E.; Houllemare, D.; Namkung, K.-C. Phenol degradation using hydroxyl radicals generated from zero-valent iron and hydrogen peroxide. *Appl. Catal. B Environ.* **2006**, *63*, 15–19. [CrossRef]

67. Kolisis, F.N.; Stamatis, H.; Flickinger, M.C. Reverse Micelles, Enzymes. In *Encyclopedia of Industrial Biotechnology*; John Wiley & Sons, Inc.: Hoboken, NJ, USA, 2009.

68. Stamatis, H.; Xenakis, A.; Kolisis, F.N. Bioorganic reactions in microemulsions: The case of lipases. *Biotechnol. Adv.* **1999**, *17*, 293–318. [CrossRef]

69. Smyth, C.; O'Driscoll, B.; Lawrence, J.; Hickey, S.; O'Reagan, T.; Buckin, V. High-Resolution Ultrasonic Spectroscopy Analysis of Microemulsions. *Pharm. Technol. Eur.* **2004**, *16*, 31–34.

70. Tonova, K.; Lazarova, Z. Reversed micelle solvents as tools of enzyme purification and enzyme-catalyzed conversion. *Biotechnol. Adv.* **2008**, *26*, 516–532. [CrossRef] [PubMed]

71. Maruhenda-Egea, F.C.; Piera-Vélasquez, S.; Cadenas, C.; Cadenas, E. Reverse micelles in organic solvents: A medium for the biotechnological use of extreme halophilic enzymes at low salt concentration. *Archaea* **2002**, *1*, 105–111. [CrossRef]

72. Dorovska-Taran, V.; Veeger, C.; Visser, A.J.W.G. Reverse micelles as a water-property-control system to investigate the hydration/activity relationship of α-chymotrypsin. *Eur. J. Biochem.* **1993**, *218*, 1013–1019. [CrossRef] [PubMed]

73. Panintrarux, C.; Adachi, S.; Araki, Y.; Kimura, Y.; Matsuno, R. Equilibrium yield of n-alkyl-β-D-glucoside through condensation of glucose and n-alcohol by β-glucosidase in a biphasic system. *Enzyme Microb. Technol.* **1995**, *17*, 32–40. [CrossRef]

74. Kouptsova, O.S.; Klyachko, N.L.; Levashov, A.V. Synthesis of Alkyl Glycosides Catalyzed by β-Glycosidases in a System of Reverse Micelles. *Russian J. Bioorganic Chem.* **2001**, *27*, 380–384. [CrossRef]

75. Grous, W.; Converse, A.; Grethlein, H.; Lynd, L. Kinetics of cellobiose hydrolysis using cellobiase composites from *Ttrichoderma reesei* and *Aspergillus niger*. *Biotechnol. Bioeng.* **1985**, *27*, 463–470. [CrossRef] [PubMed]

76. Gong, C.-S.; Ladisch, M.R.; Tsao, G.T. Cellobiase from Trichoderma viride: Purification, properties, kinetics, and mechanism. *Biotechnol. Bioeng.* **1977**, *19*, 959–981. [CrossRef] [PubMed]

77. Segel, I.H. *Enzyme Kinetics: Behavior and Analysis of Rapid Equilibrium and Steady-State Enzyme Systems*; Wiley: Hoboken, NJ, USA, 1993.

78. Boy, M.; Dominik, A.; Voss, H. Fast determination of biocatalyst process stability. *Process Biochem.* **1999**, *34*, 535–547. [CrossRef]

79. Neri, D.F.; Balcão, V.M.; Carneiro-da-Cunha, M.G.; Carvalho, L.B., Jr.; Teixeira, J.A. Immobilization of β-galactosidase from Kluyveromyces lactis onto a polysiloxane–polyvinyl alcohol magnetic (mPOS–PVA) composite for lactose hydrolysis. *Catal. Commun.* **2008**, *9*, 2334–2339. [CrossRef]

80. Dwyer, C.; Donnelly, L.; Buckin, V. Ultrasonic analysis of rennet-induced pre-gelation and gelation processes in milk. *J. Dairy Res.* **2005**, *72*, 303–310. [CrossRef] [PubMed]

81. Lehmann, L.; Buckin, V. Determination of the Heat Stability Profiles of Concentrated Milk and Milk Ingredients Using High Resolution Ultrasonic Spectroscopy. *J. Dairy Sci.* **2005**, *88*, 3121–3129. [CrossRef]

82. Griffin, W.G.; Griffin, M.C.A. The attenuation of ultrasound in aqueous suspensions of casein micelles from bovine milk. *J. Acoust. Soc. Am.* **1990**, *87*, 2541–2550. [CrossRef]

83. Holt, C.; Dalgleish, D.G. Electrophoretic and hydrodynamic properties of bovine casein micelles interpreted in terms of particles with an outer hairy layer. *J. Colloid Interface Sci.* **1986**, *114*, 513–524. [CrossRef]

84. Adler-Nissen, J. *Enzymic Hydrolysis of Food Proteins*; Elsevier Applied Science Publishers Ltd.: Amsterdam, The Netherlands, 1986.

85. Jürgens, K.D.; Baumann, R. Ultrasonic absorption studies of protein-buffer interactions. *Eur. Biophys. J.* **1985**, *12*, 217–222. [CrossRef] [PubMed]

86. Slutsky, L.; Madsen, L.; White, R. Acoustic absorption and proton-exchange kinetics in aqueous bovine pancreatic ribonuclease A. *J. Phys. Chem.* **1984**, *88*, 5679–5683. [CrossRef]

87. Slutsky, L.; Madsen, L.; White, R.; Harkness, J. Kinetics of the exchange of protons between hydrogen phosphate ions and a histidyl residue. *J. Phys. Chem.* **1980**, *84*, 1325–1329. [CrossRef]

88. Rogez, D.; Cerf, R.; Andrianjara, R.; Salehi, S.-T.; Fouladgar, H. Ultrasonic studies of proton-transfer reactions at the catalytic site of α-chymotrypsin. *FEBS Lett.* **1987**, *219*, 22–26. [CrossRef]

89. Strom-Jensen, P.R.; Dunn, F. Ultrasonic absorption by solvent–solute interactions and proton transfer in aqueous solutions of peptides and small proteins. *J. Acoust. Soc. Am.* **1984**, *75*, 960–966. [CrossRef]

90. Johnson, J.; Srisuthep, R. Physical and chemical properties of oligosaccharides. *Cereal Chem.* **1975**, *52*, 70–78.

91. Chirife, J.; Buera, M. A simple model for predicting the viscosity of sugar and oligosaccharide solutions. *J. Food Eng.* **1997**, *33*, 221–226. [CrossRef]

92. Avaltroni, F.; Bouquerand, P.; Normand, V. Maltodextrin molecular weight distribution influence on the glass transition temperature and viscosity in aqueous solutions. *Carbohydr. Polym.* **2004**, *58*, 323–334. [CrossRef]

93. Dokic, P.; Jakovljevic, J.; Dokic-Baucal, L. Molecular characteristics of maltodextrins and rheological behaviour of diluted and concentrated solutions. *Colloids Surf. A Physicochem. Eng. Asp.* **1998**, *141*, 435–440. [CrossRef]

94. Kaatze, U.; Hushcha, T.; Eggers, F. Ultrasonic Broadband Spectrometry of Liquids A Research Tool in Pure and Applied Chemistry and Chemical Physics. *J. Solut. Chem.* **2000**, *29*, 299–368. [CrossRef]

95. Epstein, P.S.; Carhart, R.R. The absorption of sound in suspensions and emulsions. I. Water fog in air. *J. Acoust. Soc. Am.* **1953**, *25*, 553–565. [CrossRef]

96. Waterman, P.C.; Truell, R. Multiple scattering of waves. *J. Math. Phys.* **1961**, *2*, 512–537. [CrossRef]

97. Fikioris, J.; Waterman, P. Multiple Scattering of Waves. II. "Hole Corrections" in the Scalar Case. *J. Math. Phys.* **1964**, *5*, 1413–1420. [CrossRef]

98. Lloyd, P.; Berry, M. Wave propagation through an assembly of spheres: IV. Relations between different multiple scattering theories. *Proc. Phys. Soc.* **1967**, *91*, 678. [CrossRef]

99. Allegra, J.; Hawley, S. Attenuation of sound in suspensions and emulsions: Theory and experiments. *J. Acoust. Soc. Am.* **1972**, *51*, 1545–1564. [CrossRef]

100. McClements, D.; Coupland, J. Theory of droplet size distribution measurements in emulsions using ultrasonic spectroscopy. *Colloids Surf. A Physicochem. Eng. Asp.* **1996**, *117*, 161–170. [CrossRef]
101. McClements, D.J.; Hemar, Y.; Herrmann, N. Incorporation of thermal overlap effects into multiple scattering theory. *J. Acoust. Soc. Am.* **1999**, *105*, 915–918. [CrossRef]

 *catalysts*

Article

Dehydrogenation of Formic Acid over a Homogeneous Ru-TPPTS Catalyst: Unwanted CO Production and Its Successful Removal by PROX

Vera Henricks [1,2], Igor Yuranov [3], Nordahl Autissier [3] and Gábor Laurenczy [1,*]

[1] Laboratory of Organometallic and Medical Chemistry, Group of Catalysis for Energy and Environment, École Polytechnique Fédérale de Lausanne, EPFL, CH-1015 Lausanne, Switzerland; v.c.henricks@student.tue.nl

[2] Molecular Catalysis, Department of Chemical Engineering and Chemistry, Eindhoven University of Technology, 5612 AZ Eindhoven, The Netherlands

[3] GRT Operations SA, CH-1350 Orbe, Switzerland; igor.iouranov@gmail.com (I.Y.); nordahl.autissier@grtgroup.swiss (N.A.)

* Correspondence: gabor.laurenczy@epfl.ch; Tel.: +41-21-693-9858

Received: 26 October 2017; Accepted: 13 November 2017; Published: 20 November 2017

Abstract: Formic acid (FA) is considered as a potential durable energy carrier. It contains ~4.4 wt % of hydrogen (or 53 g/L) which can be catalytically released and converted to electricity using a proton exchange membrane (PEM) fuel cell. Although various catalysts have been reported to be very selective towards FA dehydrogenation (resulting in H_2 and CO_2), a side-production of CO and H_2O (FA dehydration) should also be considered, because most PEM hydrogen fuel cells are poisoned by CO. In this research, a highly active aqueous catalytic system containing Ru(III) chloride and meta-trisulfonated triphenylphosphine (mTPPTS) as a ligand was applied for FA dehydrogenation in a continuous mode. CO concentration (8–70 ppm) in the resulting $H_2 + CO_2$ gas stream was measured using a wide range of reactor operating conditions. The CO concentration was found to be independent on the reactor temperature but increased with increasing FA feed. It was concluded that unwanted CO concentration in the $H_2 + CO_2$ gas stream was dependent on the current FA concentration in the reactor which was in turn dependent on the reaction design. Next, preferential oxidation (PROX) on a Pt/Al_2O_3 catalyst was applied to remove CO traces from the $H_2 + CO_2$ stream. It was demonstrated that CO concentration in the stream could be reduced to a level tolerable for PEM fuel cells (~3 ppm).

Keywords: formic acid; dehydrogenation; dehydration; ruthenium; carbon monoxide; preferential oxidation; PROX

---

1. Introduction

Currently, many methods to store and/or to transport energy in a green and durable way are being investigated in order to transition to an environmentally friendly economy. Examples include mechanical storage, batteries, superconductors, chemical storage (hydrogen and other high energy molecules) [1,2]. Formic acid (FA) containing ~4.4 wt % of H_2 has been considered a potential energy carrier since 1978 [3]. Essentially, H_2 could be chemically combined with carbon dioxide (CO_2) to form liquid FA [4], which is more easily stored and transported than gaseous H_2 considering practical and safety issues. FA can then be catalytically decomposed back to H_2 and CO_2 in a 1:1 ratio, closing a loop for CO_2 and providing H_2 for fuel cell technology to generate electricity. Thus, FA would simply function as an intermediate to store and to transport energy in a clean way (Figure 1).

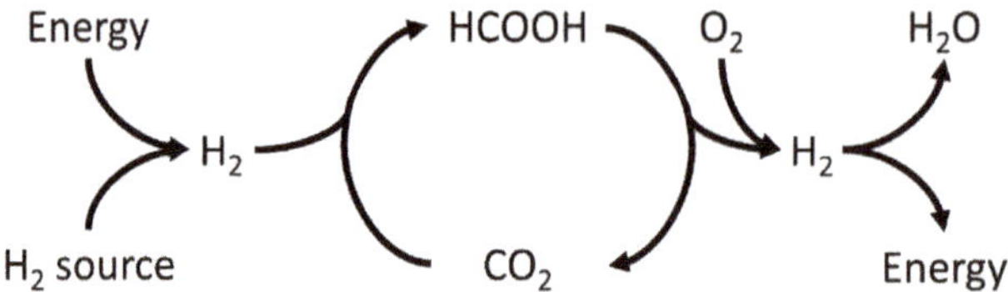

Figure 1. Formic acid as an energy carrier providing a closed loop for CO_2.

Recently, FA has sparked interest again, because catalysts highly effective in FA dehydrogenation in reasonable conditions have been found [5–9]. This fact in combination with easy and safe handling opens a route of using FA in applications such as energy storage and transport [10].

Although very high catalyst selectivities have been reported in literature [7,9,11], a side-reaction of FA dehydration takes usually place, next to FA dehydrogenation, due to a still insufficient catalyst selectivity and/or thermal decomposition of FA:

$$HCOOH \rightarrow H_2 + CO_2, \tag{1}$$

$$HCOOH \rightarrow H_2O + CO. \tag{2}$$

It is very important to consider this side-reaction, because a CO content, even at ppm-levels, is poisonous for many fuel cells [12,13]. FA thermal decomposition was investigated both computationally and experimentally. However, most of the experiments were conducted in supercritical water [14–17].

Mitigating strategies such as improving CO-tolerance of fuel cells showed that Pt–Ru alloying, for example, can weaken the binding of CO to Pt catalysts, decreasing the poisoning effect [18,19]. Air-bleed—injecting a small amount of air or oxygen into the fuel feed stream—is also applied to lessen CO poisoning. Here, CO oxidation takes place directly on a fuel cell membrane as an oxidation catalyst. The method can recover up to 90% of the fuel cell performance at 200 ppm CO [20,21]. However, air-bleed was shown to cause long-term degradation of the membrane where O_2 is reduced to H_2O_2 which corrodes the catalyst [22]. To avoid this problem, preferential oxidation (PROX) could be used to decrease CO concentration. In this approach, CO is removed from the H_2 stream through injecting a small amount of O_2 and selective CO oxidation over a heterogeneous catalyst prior to entering the fuel cell. Nowadays, PROX is mainly applied to bulk production of H_2 where a significant CO amount (0.5–1%) is produced through steam reforming and water gas shift reactions [23]. Several catalysts such as noble metals supported on alumina or ceria as well as transition metals on similar supports are available [24]. The catalysts are designed to work in gas streams containing at least 85% of H_2. In H_2 rich gas streams, Equations (3) and (4) are of importance

$$2CO + O_2 \rightarrow 2CO_2 \tag{3}$$

$$2H_2 + O_2 \rightarrow 2H_2O \tag{4}$$

To minimize H_2 loss and H_2O emittance (Equation (4)), the catalyst selectivity towards CO oxidation needs to be as high as possible. An important factor is the amount of supplied O_2, because selective catalysts start converting H_2 after complete CO conversion [24]. Moreover, FA dehydrogenation gas mixtures contain much more CO_2 (~50%) which could possibly effect the conversion of CO.

Other methods to remove CO from gas streams are adsorption and methanation. However, it is hard to implement pressure swing CO adsorption in compact applications [25]. Methanation is usually mentioned as an option next to PROX:

$$CO + 3H_2 \rightarrow CH_4 + H_2O, \tag{5}$$

$$CO_2 + 4H_2 \rightarrow CH_4 + 2H_2O. \tag{6}$$

Supported Ni and Ru methanation catalysts are well investigated [26]. Fuel cells are tolerant to methane [12]. No gas needs to be added to the stream. However, significant amount of H_2 could be lost in this process depending on the CO concentration and catalyst selectivity towards CO methanation. Next to that, most of the methanation catalysts become active only at temperatures over 250 °C.

In the present study, a $H_2 + CO_2$ gas mixture (reformate) resulted from catalytic FA dehydrogenation carried out in a continuous-flow gas–liquid bed reactor was analyzed. The liquid reaction phase was comprised of an aqueous solution of Ru(III) chloride and *m*TPPTS used as catalyst precursors. *m*TPPTS was chosen as a phosphine-ligand because of its high stability and water solubility of the resulting catalytically active complex [7]. Previously, it was reported that this catalyst was stable over more than one year of intermittent use, while being kept in air [27]. The turnover frequencies (TOF's) reached in a continuous mode were 210 h^{-1} and 670 h^{-1} at 100 °C and 120 °C, respectively [27]. CO content in the resulting gas mixture was not measured in a continuous mode, but batch experiments performed using the same reactant concentrations showed no CO traces (detection limit ~3 ppm) [7].

In the present study, FA was fed to the reactor continuously. The CO concentration (ppm level) in the $H_2 + CO_2$ gas streams was measured as a function of the reaction temperature and FA feed flow. Quantification of such low CO concentrations (especially in CO_2-rich gas mixtures) is an extremely difficult task. In this research, gas chromatography (GC) coupled with sample methanation was used to achieve a good CO/CO_2 separation and sensibility of the measurements, while it was also possible to measure a H_2 concentration.

Next, CO removal from $H_2 + CO_2$ gas streams by PROX was tested using a commercially available Pt/Al_2O_3 catalyst placed downstream from the FA dehydrogenation reactor. The PROX catalyst was not heated. A controlled amount of O_2 (in air) was injected into the gas stream before the PROX catalyst. The considered parameters were the CO concentration in the gas stream and the amount of added O_2. A combination of these numbers can be expressed through the O_2 excess parameter λ (Equation (7)), representing the O_2 excess, relatively to its amount required for total CO oxidation. At $\lambda = 1$, the O_2 concentration in a gas mixture is exactly enough to convert all CO (catalyst selectivity ~100%)

$$\lambda = \frac{2C_{O_2}}{C_{CO}} = \frac{2p_{O_2}}{p_{CO}} \tag{7}$$

2. Results and Discussion

2.1. CO Side-Production in FA Dehydrogenation

The CO concentration in $H_2 + CO_2$ gas mixtures was measured in the reaction temperature range of 60–80 °C. Below 60 °C, the Pt/Al_2O_3 catalyst activity was too low for the application, while at 80 °C and higher, water evaporation became quite intensive constantly reducing the catalyst solution volume. At a steady-state, the total gas flow rate was found to be the same for the all measured temperatures (Table 1) indicating that in this temperature range the reaction rate was limited by the FA feed flow rate.

Table 1. Total $H_2 + CO_2$ gas flow as a function of FA feed measured at 60, 70 and 80 °C.

FA Feed (mL/min)	Total Gas Flow (L/min)
0.3	0.43
0.6	0.86
0.9	1.29

As can be seen in Figure 2, the CO concentration in the gas mixture was below 10 ppm at the FA feed flow rate of 0.3 mL/min but increased with increasing FA feed. It was supposed that, due to an imperfect reactant mixing in the reactor, at the high FA feed flow rates the size of injected FA droplets is bigger and hence their lifetime is longer. At higher temperatures, it leads to more pronounced side-reaction of FA thermal dehydration and higher CO content in the reformate.

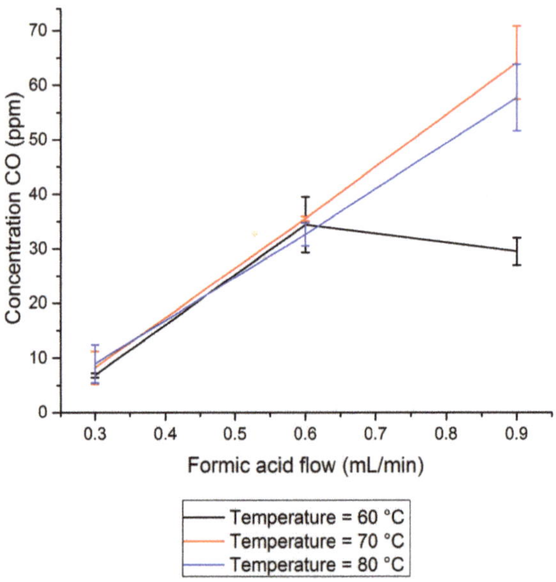

Figure 2. CO concentration in $H_2 + CO_2$ gas mixtures as a function of FA feed flow and temperature.

2.2. PROX

Figure 3 shows CO concentrations in $H_2 + CO_2$ gas mixtures treated on the PROX catalyst. Varying the FA feed flow at constant reaction temperature, both the total gas flow rate and CO concentration were varied. It was observed during the experiments that the catalyst container became warm (50–60 °C) due to the exothermic oxidation of CO and H_2. It was found for all reformate flow rates that at high concentrations of injected O_2 ($\lambda = 3$–14), the PROX treatment led to a drop in the CO concentrations to ~3 ppm. At $\lambda = 1$–2 a small increase of the resulting CO concentration up to 4–6 ppm was observed. Surprisingly, at $\lambda < 1$, a high CO conversion was observed as well (Figure 3). This fact was explained by the presence of a considerable amount of O_2 adsorbed on the catalyst surface. After an N_2 purge of the PROX catalyst for three days, the CO concentration decreased under the same conditions ($\lambda < 1$) from initial 35 to 16 ppm indicating O_2 strongly adsorbed on the catalyst that could not be removed by a simple purge.

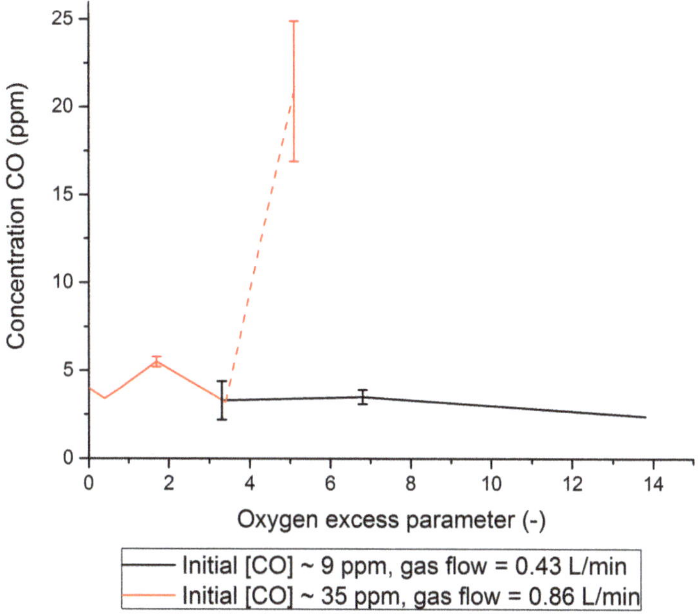

Figure 3. CO concentration after PROX oxidation in H_2 + CO_2 gas mixtures as a function of initial CO concentration, total flow rate, and injected O_2.

It is worth noticing that at the total H_2 + CO_2 gas flow rate of 0.86 L/min and high O_2 concentrations ($\lambda > 5$), water condensation in the PROX catalyst container and drop of the CO conversion were detected (Figure 3) indicating a blockage of the catalyst surface by water.

This effect was previously reported for PROX catalysts [24]. For the lower gas flow rates, water clogging was observed only after a long period of catalyst exposure. The activity of the catalyst deactivated by the water clogging could be completely recovered by an air or nitrogen purge at room temperature.

3. Materials and Methods

A homogeneous Ru-*m*TPPTS catalyst (*m*TPPTS: meta-trisulfonated triphenylphosphine, Figure 4) solution (33 mM, 700 mL) was prepared by combining an aqueous solution of $RuCl_3$ (5.6 g) and *m*TPPTS (2.8 equivalent) with FA (128 g) and Na formate (30 g). The catalyst was activated prior to its placing into the reactor by steering the solution at 60 °C for 4 h under a N_2 flow.

Figure 4. TPPTS.

FA was dehydrogenated in a 1 L continuous-flow gas–liquid bed reactor (Figure 5). A cylindrical, non-stirred reactor vessel of a 7.0 cm inner diameter and 40 cm length was heated by means of an internal heating element connecting to an oil circulator (Huber Pilot One). The temperature inside the reactor was varied in the range of 60–80 °C. FA was injected into the reactor (0.3–0.9 mL/min) at the reactor bottom through a HPLC pump (HPLC Pump 422, Kontron Instruments, Montigny Le Bretonneux, France). The gas bubbles produced due to the reaction rose, mixing the reaction solution. Outside the reactor, a $H_2 + CO_2$ gas mixture was passed through a condenser and an adsorber (activated carbon) to remove FA and water vapors. The total gas flow was measured using a bubble flowmeter. In separate experiments, a $H_2 + CO_2$ gas mixture was passed through a container of 2.8 cm inner diameter, containing a PROX catalyst (50 g). A commercial 0.5 wt % Pt supported on alumina (spheres of 1.8–3.5 mm diameter, Johnson Matthey, Royston, UK) was used as a PROX catalyst. An air flow controlled by a mass flow controller was injected into the gas stream before the PROX catalyst.

Gas samples were collected at a steady state before and after the PROX catalyst using a rubber balloon flushed with nitrogen and vacuumed. The collected gas samples were analyzed by a GC method using an Agilent 7890B gas chromatograph (Agilent, Santa Clara, CA, USA) equipped with a CarboPlot P7 column (25 m × 0.53 mm × 25 μm), a Nickel Catalyst Kit (to convert CO (CO_2) to CH_4), flame ionization (FID), and thermal conductivity (TCD) detectors. At least three samples were taken for every point. At least three GC measurements were done for every gas sample. The GC was calibrated using standard (50% H_2 + 40 ppm CO + CO_2) and (50% CO_2 + 100 ppm CO + N_2) mixtures (Carbagas, Lausanne, Switzerland). The detection limit of the GC method was ~3 ppm.

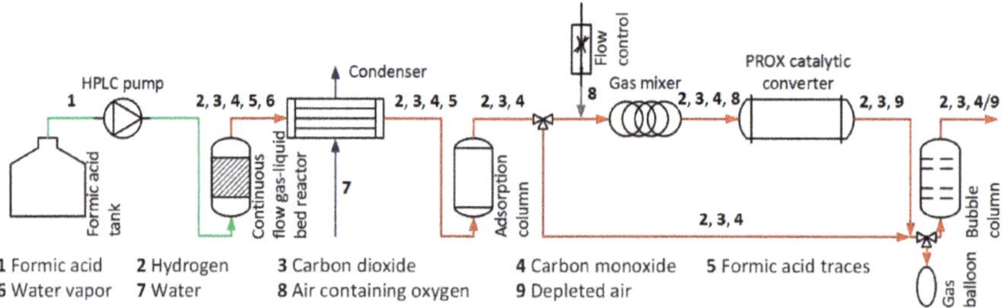

Figure 5. Scheme of the setup.

4. Conclusions

1. It was found that unwanted CO concentrations in $H_2 + CO_2$ gas streams catalytically produced through FA dehydrogenation over a homogeneous Ru-*m*TPPTS catalyst in a continuous-flow gas–liquid bed reactor (60–80 °C) was too high (8–70 ppm) to be used directly in a fuel cell. The CO concentration was dependent on the FA feed flow rate and reaction design (mixing).

2. It was shown that the Pt/Al_2O_3 catalyst can be easily implemented to clean $H_2 + CO_2$ gas streams from CO by PROX. It was shown that the O_2 supply should be fine-tuned well to prevent both a water clogging of the PROX catalyst and a H_2 loss due to its oxidation. Experimentally, CO concentration was confidently decreased by PROX at O_2 concentrations close to stoichiometry (0.008–0.065 v % O_2) to the level acceptable for fuel cell applications (<5 ppm). To our knowledge, this is the first example of successful CO removal from a FA reformate by PROX.

Acknowledgments: École Polytechnique Fédérale de Lausanne (EPFL), Swiss Competence Center for Energy Research (SCCER), Swiss Commission for Technology and Innovation (CTI) and Swiss National Science Foundation (SNSF) are thanked for financial support.

Author Contributions: G.L., I.Y. and V.H. conceived and designed the experiments; V.H. performed the measurements; N.A. and G.L. discussed the results, V.H., I.Y. and G.L. analyzed the data; V.H., I.Y. and G.L. wrote the paper.

Conflicts of Interest: The authors declare no conflict of interest.

References

1. Zhu, Q.-L.; Xu, Q. Liquid Organic and Inorganic Chemical Hydrides for High-Capacity Hydrogen Storage. *Energy Environ. Sci.* **2015**, *8*, 478–512. [CrossRef]
2. Dalebrook, A.F.; Gan, W.; Grasemann, M.; Moret, S.; Laurenczy, G. Hydrogen Storage: Beyond Conventional Methods. *Chem. Commun.* **2013**, *49*, 8735–8751. [CrossRef] [PubMed]
3. Sordakis, K.; Tang, C.; Vogt, L.K.; Junge, H.; Dyson, P.J.; Beller, M.; Laurenczy, G. Homogeneous Catalysis for Sustainable Hydrogen Storage in Formic Acid and Alcohols. *Chem. Rev.* **2017**. [CrossRef] [PubMed]
4. Moret, S.; Dyson, P.J.; Laurenczy, G. Direct Synthesis of Formic Acid from Carbon Dioxide by Hydrogenation in Acidic Media. *Nat. Commun.* **2014**, *5*, 4017. [CrossRef] [PubMed]
5. Eppinger, J.; Huang, K.-W. Formic Acid as a Hydrogen Energy Carrier. *ACS Energy Lett.* **2017**, *2*, 188–195. [CrossRef]
6. Li, Z.; Xu, Q. Metal-Nanoparticle-Catalyzed Hydrogen Generation from Formic Acid. *Acc. Chem. Res.* **2017**, *50*, 1449–1458. [CrossRef] [PubMed]
7. Fellay, C.; Dyson, P.J.; Laurenczy, G. A viable hydrogen-storage system based on selective formic acid decomposition with a ruthenium catalyst. *Angew. Chem. Int. Ed.* **2008**, *47*, 3966–3968. [CrossRef] [PubMed]
8. Boddien, A.; Mellmann, D.; Gärtner, F.; Jackstell, R.; Junge, H.; Dyson, P.J.; Laurenczy, G.; Ludwig, R.; Beller, M. Efficient Dehydrogenation of Formic Acid Using an Iron Catalyst. *Science* **2011**, *333*, 1733–1736. [CrossRef] [PubMed]
9. Himeda, Y.; Miyazawa, S.; Hirose, T. Interconversion between Formic Acid and H_2/CO_2 Using Rhodium and Ruthenium Catalysts for CO_2 Fixation and H_2 Storage. *ChemSusChem* **2011**, *4*, 487–493. [CrossRef] [PubMed]
10. Hull, J.F.; Himeda, Y.; Wang, W.-H.; Hashiguchi, B.; Periana, R.; Szalda, D.J.; Muckerman, J.T.; Fujita, E. Reversible Hydrogen Storage Using CO_2 and a Proton-Switchable Iridium Catalyst in Aqueous Media under Mild Temperatures and Pressures. *Nat. Chem.* **2012**, *4*, 383–388. [CrossRef] [PubMed]
11. Gan, W.; Dyson, P.J.; Laurenczy, G. Heterogeneous silica-supported ruthenium phosphine catalysts for selective formic acid decomposition. *ChemCatChem* **2013**, *5*, 3124–3130. [CrossRef]
12. Sharaf, O.Z.; Orhan, M.F. An overview of fuel cell technology: Fundamentals and applications. *Renew. Sustain. Energy Rev.* **2014**, *32*, 810–853. [CrossRef]
13. Baschuk, J.J.; Li, X. Modelling CO poisoning and O_2 bleeding in a PEM fuel cell anode. *Int. J. Energy Res.* **2003**, *27*, 1095–1116. [CrossRef]
14. Bröll, D.; Kaul, C.; Krämer, A.; Krammer, P.; Richter, T.; Jung, M.; Vogel, H.; Zehner, P. Chemistry in Supercritical Water. *Angew. Chem. Int. Ed. Engl.* **1999**, *38*, 2998–3014. [CrossRef]
15. Wakai, C.; Yoshida, K.; Tsujino, Y.; Matubayasi, N.; Nakahara, M. Effect of Concentration, Acid, Temperature, and Metal on Competitive Reaction Pathways for Decarbonylation and Decarboxylation of Formic Acid in Hot Water. *Chem. Lett.* **2004**, *33*, 572–573. [CrossRef]
16. Bjerre, A.B.; Sorensen, E. Thermal Decomposition of Dilute Aquaous Formic Acid Solutions. *Ind. Eng. Chem. Res.* **1992**, *31*, 1574–1577. [CrossRef]
17. Akiya, N.; Savage, P.E. Role of water in formic acid decomposition. *AIChE J.* **1998**, *44*, 405–415. [CrossRef]
18. Koper, M.T.M.; Shubina, T.E.; Van Santen, R.A. Periodic Density Functional Study of CO and OH Adsorption on Pt–Ru Alloy Surfaces: Implications for CO Tolerant Fuel Cell Catalysts. *J. Phys. Chem. B* **2002**, *106*, 686–692. [CrossRef]
19. Yano, H.; Ono, C.; Shiroishi, H.; Okada, T. New CO tolerant electro-catalysts exceeding Pt–Ru for the anode of fuel cells. *Chem. Commun.* **2005**, 1212–1214. [CrossRef] [PubMed]
20. Sung, L.Y.; Hwang, B.J.; Hsueh, K.L.; Tsau, F.H. Effects of anode air bleeding on the performance of CO-poisoned proton-exchange membrane fuel cells. *J. Power Sources* **2010**, *195*, 1630–1639. [CrossRef]
21. Sung, L.-Y.; Hwang, B.-J.; Hsueh, K.-L.; Su, W.-N.; Yang, C.-C. Comprehensive study of an air bleeding technique on the performance of a proton-exchange membrane fuel cell subjected to CO poisoning. *J. Power Sources* **2013**, *242*, 264–272. [CrossRef]
22. Schmittinger, W.; Vahidi, A. A review of the main parameters influencing long-term performance and durability of PEM fuel cells. *J. Power Sources* **2008**, *180*, 1–14. [CrossRef]

23. Mariño, F.; Descorme, C.; Duprez, D. Noble metal catalysts for the preferential oxidation of carbon monoxide in the presence of hydrogen (PROX). *Appl. Catal. B Environ.* **2004**, *54*, 59–66. [CrossRef]
24. Bion, N.; Epron, F.; Moreno, M.; Mariño, F.; Duprez, D. Preferential oxidation of carbon monoxide in the presence of hydrogen (PROX) over noble metals and transition metal oxides: Advantages and drawbacks. *Top. Catal.* **2008**, *51*, 76–88. [CrossRef]
25. Formanski, V.; Kalk, T.; Roes, J. Compact hydrogen production systems for solid polymer fuel cells. *J. Power Sources* **1998**, *71*, 199–207.
26. Panagiotopoulou, P.; Kondarides, D.I.; Verykios, X.E. Selective methanation of CO over supported noble metal catalysts: Effects of the nature of the metallic phase on catalytic performance. *Appl. Catal. A Gen.* **2008**, *344*, 45–54. [CrossRef]
27. Fellay, C.; Yan, N.; Dyson, P.J.; Laurenczy, G. Selective formic acid decomposition for high-pressure hydrogen generation: A mechanistic study. *Chem. Eur. J.* **2009**, *15*, 3752–3760. [CrossRef] [PubMed]

 *catalysts*

Article

Catalytic Activities of Ribozymes and DNAzymes in Water and Mixed Aqueous Media

Shu-ichi Nakano [1,*], **Masao Horita** [1], **Miku Kobayashi** [1] and **Naoki Sugimoto** [1,2]

[1] Department of Nanobiochemistry, Faculty of Frontiers of Innovative Research in Science and Technology (FIRST), Konan University, 7-1-20, Minatojima-Minamimachi, Chuo-ku, Kobe 650-0047, Japan; s1491036@s.konan-u.ac.jp (M.H.); miiiku0819@gmail.com (M.K.); sugimoto@konan-u.ac.jp (N.S.)

[2] Frontier Institute for Biomolecular Engineering Research (FIBER), Konan University, 7-1-20, Minatojima-Minamimachi, Chuo-ku, Kobe 650-0047, Japan

* Correspondence: shuichi@center.konan-u.ac.jp; Tel.: +81-78-303-1429

Received: 1 November 2017; Accepted: 20 November 2017; Published: 23 November 2017

Abstract: Catalytic nucleic acids are regarded as potential therapeutic agents and biosensors. The catalytic activities of nucleic acid enzymes are usually investigated in dilute aqueous solutions, although the physical properties of the reaction environment inside living cells and that in the area proximal to the surface of biosensors in which they operate are quite different from those of pure water. The effect of the molecular environment is also an important focus of research aimed at improving and expanding nucleic acid function by addition of organic solvents to aqueous solutions. In this study, the catalytic activities of RNA and DNA enzymes (hammerhead ribozyme, 17E DNAzyme, R3C ribozyme, and 9DB1 DNAzyme) were investigated using 21 different mixed aqueous solutions comprising organic compounds. Kinetic measurements indicated that these enzymes can display enhanced catalytic activity in mixed solutions with respect to the solution containing no organic additives. Correlation analyses revealed that the turnover rate of the reaction catalyzed by hammerhead ribozyme increased in a medium with a lower dielectric constant than water, and the turnover rate of the reaction catalyzed by 17E DNAzyme increased in conditions that increased the strength of DNA interactions. On the other hand, R3C ribozyme and 9DB1 DNAzyme displayed no significant turnover activity, but their single-turnover rates increased in many mixed solutions. Our data provide insight into the activity of catalytic nucleic acids under various conditions that are applicable to the medical and technology fields, such as in living cells and in biosensors.

Keywords: ribozyme; DNAzyme; rate constant; melting temperature; molecular crowding; polyethylene glycol

---

1. Introduction

The catalytic activities of RNA enzymes (ribozymes) and DNA enzymes (DNAzymes) are useful in the development of medical and biotechnology tools. Several types of nucleic acid enzymes have been found in living cells or were artificially made through in vitro selection using libraries of random sequence [1]. Specifically, the hammerhead ribozyme and the 8–17 DNAzyme, which effectively catalyze the site-specific cleavage of the phosphodiester bond of an RNA substrate, have been used as inhibitors of intracellular gene expression [2,3] and as biosensors for detecting target metal ions, molecules, or specific DNA and RNA sequences [4–9]. The flexibility in nucleic acid sequence design enables to convert these enzymes into ligand-responsive aptazyme systems regulating gene expression levels and into sensitive electrochemical sensors [10–13].

Nucleic acid enzymes are highly negatively charged, so they exhibit catalytic activity in the presence of metal ions, like Mg^{2+}. In general, cation binding is required for the hybridization of the enzyme with its substrate strand, for the formation of a catalytically active complex, and for the catalytic chemical process to take place [14–19]. Studies using simple aqueous solutions have provided detailed information regarding the catalytic activity of many types of ribozymes and DNAzymes and the metal ion conditions necessary for catalysis. However, the environment inside living cells and that in the area proximal to the surface of biosensors, where nucleic acid enzymes exert their therapeutic and diagnostic activity, are quite different from the aqueous solutions commonly used for in vitro evaluation of catalytic activity. In these particular cases, nucleic acid interactions occur under sterically-restricted conditions that alter the diffusion, dynamics, and effective concentration of molecules. Moreover, intracellular water and that in the vicinity of the surface of a biosensor have different physical properties (e.g., dielectric constant and water activity) from pure water. The characteristics of these environments are known to affect the thermodynamics and kinetics of nucleic acid interactions [20]. The molecular environment of an aqueous solution is also substantially affected by addition of organic solvents. Solvents like ethanol, methanol, dimethyl sulfoxide (DMSO), and *N,N*-dimethylformamide (DMF) have been employed to improve oligonucleotide function in the polymerase chain reaction [21,22], molecular beacon assays [23], DNA strand exchanges [24], and to assist DNAzyme functionalization [25–28]. The presence of organic solvents changes the physical properties of a solution, and the organic molecules themselves may interact with nucleic acids. Understanding the effects of the molecular environment on the activity of nucleic acid enzymes is helpful for the use of the enzymes in practical applications.

Aqueous solutions containing water-soluble organic compounds at a concentration of a few to several tenths of a percent have been used to investigate how environmental factors influence nucleic acid behavior. These types of studies have shown that the solution composition influences base-pair interaction energy, RNA folding, hydration status, and nucleic acid–metal ion interactions [20,29]. Polyethylene glycol (PEG) is a commonly used reagent in these studies. This polymer is water-soluble, and highly purified PEGs of different molecular weights are commercially available. Notably, low-molecular-weight PEGs exist as liquids and high-molecular-weight PEGs as solids at room temperature. Mixed solutions, PEG-containing solutions in many cases, have been reported to increase the rate of RNA cleavage by ribozymes, such as the group I intron ribozyme, human delta virus-like ribozyme, hairpin ribozyme, leadzyme, and hammerhead ribozyme (see the references in [30]). Several reports also exist on the catalytic activity of DNAzymes; in these cases, the reaction rates in mixed solutions containing ethanol or other organic solvents were higher than those in the solution containing no organic solvents [26–28]. These studies suggest the possibility of using mixed solutions to enhance the activity of nucleic acid enzymes. However, the effects of solution composition on nucleic acid behavior have been explained by a variety of different environmental factors: excluded volume interactions (molecular crowding environment), reduced water activity (dehydrating environment), and reduced dielectric constant (stronger electrostatic interaction environment) [31]. This uncertainty is partly the result of the limited use of mixed solutions to discuss the major environmental factor in these studies. In addition, most of the reported studies evaluated the catalytic activity under single-turnover conditions. In contrast to the case of protein enzymes, the turnover of nucleic acid enzymes is intrinsically difficult to achieve, because of the stable base-pairing interactions that are established with substrate strands and because of the slow kinetics of Watson-Crick base-pair opening. Methods for enhancing the turnover number would be useful to increase the effectiveness of technologies based on the catalytic activity of nucleic acid enzymes.

The hypothesis can be made that organic solvents that reduce the strength of base-pairing interactions facilitate the release of reaction products and thus increase the catalytic turnover of the reactions catalyzed by ribozymes and DNAzymes. Here, we investigated the activities of several types of RNA-cleaving and RNA-ligating nucleic acid enzymes under single- and multiple-turnover conditions, using 21 different mixed aqueous solutions containing PEGs with an average molecular weight ranging from 2×10^4 to 2×10^2, ethylene glycol derivatives (ethylene glycol (EG), glycerol (Glyc), 1,3-propanediol (PDO), 2-methoxyethanol (MME), and 1,2-dimethoxyethane (DME)), small primary alcohols (methanol (MeOH), ethanol (EtOH), and 1-propanol (PrOH)), amide compounds (urea, formamide (FA), DMF, and acetamide (AcAm)), aprotic compounds (acetonitrile (AcCN), DMSO, and 1,4-dioxane (DOX)), or dextran with an average molecular weight of 1×10^4 (Dex). These solutions have different dielectric constant, water activity, and viscosity. Furthermore, the additives may directly interact with nucleic acids, and additives of large size occupy large volumes from which other molecules are excluded. These factors potentially influence the thermodynamics and kinetics of nucleic acid interactions [20]. Results show that the ribozymes and DNAzymes used in this study remain active in many mixed solutions, and they may even display enhanced catalytic activity. Although destabilization of base pairing between the enzyme and substrate was expected to beneficial for the catalytic turnover, the mixed solutions did not increase the turnover rate of the enzymes having extended base pairing. Our data provide insight into the catalytic activity of nucleic acid enzymes in the presence of organic compounds. This insight can improve the quality of medical and technological applications of nucleic acid enzymes, such as their use in living cells or as biosensors.

2. Results

2.1. Substrate Cleavage by Ribozymes

The hammerhead ribozyme is a small RNA-cleaving ribozyme found in nature. Because of its small size and high nuclease activity, this ribozyme is an attractive candidate for therapeutic and diagnostic uses [2]. **HH(S)** and **HH(L)** investigated in this study are hammerhead ribozymes having different number of base pairs formed with the substrate RNA (Figure 1a). **HH(S)** is a small sequence motif that can form the catalytic core. **HH(L)** has the extended base pairing that allows the ribozyme to form loop-loop tertiary interactions required for fast cleavage of the substrate [32,33]. Both ribozymes exhibited high RNA cleavage activity under single-turnover condition in the presence of 10 mM $MgCl_2$, but the kinetic traces associated with substrate cleavage differed for the two systems. The time course of substrate cleavage by **HH(S)** was closely approximated by a single-exponential function [34], whereas **HH(L)** showed a biphasic kinetic behavior, whereby ~10% of the substrate was cleaved at a fast rate (within less than a few seconds) and ~70% of it was cleaved at a slow rate (taking a few tens of minutes), so that 80% substrate conversion was achieved after completion of the slow phase of the reaction (Figure 1b). This observation is consistent with a previous report that an extended ribozyme having a similar sequence to **HH(L)** showed a biphasic kinetic behavior [32]. It is regarded that substrate cleavage by the extended ribozyme proceeds through multiple reaction pathways: the fast phase of the reaction reflects the cleavage of the catalytically active complex, whereas its slow phase reflects the contribution of the slow conversion of kinetically trapped conformations, such as those due to misfolding or interfering secondary structures, to the active complex.

The presence of 20% PEG8000 (with an average molecular weight of 8×10^3) or PEG200 (2×10^2) influenced the single-turnover reaction rate. We have previously reported the enhancement of the rate of cleavage catalyzed by **HH(S)** in mixed solutions containing PEG8000 or PEG200 [34,35]. When **HH(L)** was investigated, the effects of PEG on the fast and slow phases of the reaction were different: the amplitude of the fast phase conversion increased 9.8-fold with PEG8000 and 5.3-fold with PEG200, although the rates were too fast for the rate constants to be determined. The rate of the slow phase was not affected by the addition of PEGs, and the amplitude after completion of the slow phase reached ~80% conversion, the same level observed in the absence of PEG (Figure 1b). These results suggest that the molecular environment made up of PEG-containing aqueous solutions increases the amount of the catalytically active complex with respect to the case of the solution containing no organic additives; however, it does not significantly affect the rate of the conformational change necessary for the formation of the active complex.

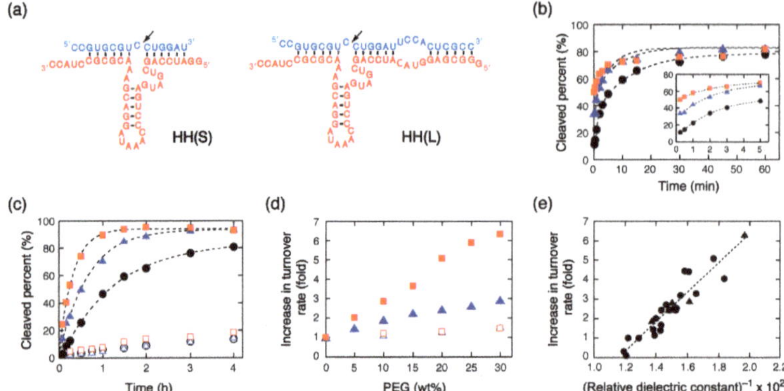

Figure 1. (**a**) Structures of the hammerhead ribozymes and their substrates used in this study. The cleavage sites are indicated by arrowheads. (**b**) Kinetic traces for substrate cleavage catalyzed by **HH(L)** under single-turnover conditions in the presence of 10 mM $MgCl_2$ and in the absence (black circles) and presence of PEG8000 (red squares) or PEG200 (blue triangles). The data relative to the first 5 min of the reaction are enlarged in the inset. (**c**) Kinetic traces for substrate cleavage catalyzed by **HH(S)** (closed symbols) and **HH(L)** (open symbols) under multiple-turnover conditions in the presence of 10 mM $MgCl_2$ and in the absence (black circles) and presence of PEG8000 (red squares) or PEG200 (blue triangles). (**d**) Increments of the turnover rate of the reactions catalyzed by **HH(S)** (closed symbols) or **HH(L)** (open symbols) in the presence of 10 mM $MgCl_2$ plotted against the amount of PEG8000 (red squares) or PEG200 (blue triangles) in solution. (**e**) Plot of the increments of the turnover rate of the **HH(S)**-catalyzed reaction in the presence of 10 mM $MgCl_2$ against the inverse of the relative dielectric constant of solutions. The data obtained using 10% or 30% PEG solutions (PEG8000 and PEG200) are also included (triangles). The correlation coefficient of a linear fit is 0.94.

The cleavage rates of **HH(S)** and **HH(L)** under multiple-turnover conditions were investigated using a 20-fold excess of substrate RNA to ribozyme. In principle, at least two kinetic stages exist for the reaction: the first round of substrate cleavage (accounting at most for 5% of the reaction yield) and the subsequent rounds of cleavage. The kinetic data showed that **HH(S)** had high turnover activity in the presence of 10 mM $MgCl_2$ (the turnover number was greater than 10 after 1.5 h), and the turnover rate increased 5.1-fold after addition of PEG8000 and 2.4-fold after addition of PEG200 (Figure 1c). By contrast, the extended ribozyme **HH(L)** displayed lower turnover activity than **HH(S)**, both in the presence and absence of PEGs (the turnover number was less than three after 6 h). The turnover rate of the **HH(S)**-catalyzed reaction increased linearly with the amount of PEG (5–30%), but the turnover rate of the **HH(L)**-catalyzed reaction did not change significantly even when using 30% PEG solutions (Figure 1d).

To identify the environmental factor causing the observed effect of PEGs on **HH(S)**, we investigated the effects of other mixed solutions having various physical properties. The kinetic data showed that the turnover rate increased in many mixed solutions with respect to the solution containing no organic additives (Table 1). Analysis of the correlation with the values of solution properties revealed that the turnover rate was highly correlated with the dielectric constant of the medium (Figure 1e), but the rate was not correlated with water activity, viscosity, or the additives' molecular weight (Figure S1). These results suggest that the rate enhancement in mixed solutions was predominantly caused by electrostatic effects. These same correlations were previously reported for the single-turnover rates [35], suggesting that reaction rate enhancements resulted from the same mechanism in both reaction conditions. The decrease in dielectric constant observed as a result of the formation of mixed solutions may strengthen the binding interaction between nucleic acids and Mg^{2+} ions, which is a required process for the formation of the active complex and for catalysis. Consistent with the strengthening of Mg^{2+} binding to the ribozyme, the increase in turnover rate was more pronounced when $MgCl_2$ concentration decreased to 1 mM. In particular, the turnover rate increased more than 10-fold in mixed solutions containing PEG20000, PEG8000, and PEG2000 (13, 17, and 10-fold, respectively) with respect to the solution containing no organic additives (Table 1).

Table 1. Fold increase in the rate of the **HH(S)**-catalyzed turnover reaction in mixed solutions over the rate in the absence of organic additives.

Solution [1]	10 mM $MgCl_2$	1 mM $MgCl_2$	Solution [1]	10 mM $MgCl_2$	1 mM $MgCl_2$	Solution [1]	10 mM $MgCl_2$	1 mM $MgCl_2$
PEG20000	3.4	13	PDO	1.7	2.8	FA	0.32	<0.02
PEG8000	5.1	17	MME	2.4	3.9	DMF	1.3	<0.02
PEG2000	3.3	10	DME	4.4	11	AcAm	1.0	<0.02
PEG600	2.5	6.8	MeOH	2.4	4.9	AcCN	2.7	6.5
PEG200	2.4	5.5	EtOH	2.5	4.8	DMSO	1.4	1.2
EG	2.0	1.3	PrOH	3.2	6.0	DOX	1.9	4.6
Glyc	1.1	1.3	Urea	0.10	<0.02	Dex	1.4	2.5

[1] Abbreviations: PEG (polyethylene glycol with an average molecular weight ranging from 2×10^4 to 2×10^2); EG (ethylene glycol); Glyc (glycerol); POD (1,3-propanediol); MME (2-methoxyethanol); DME (1,2-dimethoxyethane); MeOH (methanol); EtOH (ethanol); PrOH (1-propanol); FA (formamide); DMF (*N,N*-dimethylformamide); AcAm (acetamide); AcCN (acetonitrile); DMSO (dimethyl sulfoxide); DOX (1,4-dioxane); and Dex (dextran with an average molecular weight of 1×10^4).

2.2. Substrate Cleavage by DNAzymes

The DNA enzyme named 17E DNAzyme is a variant of the 8–17 DNAzyme obtained through in vitro selection [36,37]. This enzyme is relatively small in size and can cleave a substrate sequence in the presence of Mg^{2+} or other metal ions. Notably, it can be engineered to have nuclease activity in various situations, including in vivo, in blood serum, and when anchored on the surface of a biosensor [7,38–40]. Notably, this enzyme is being tested in clinical trials [3]. **17E(S)** and **17E(L)** are DNAzymes having different number of base pairs formed with the substrate DNA–RNA chimeric strand, and substrate hydrolysis occurs at a single RNA nucleotide (rA) site (Figure 2a). With these DNAzymes, rapid substrate cleavage was observed, reaching ~90% efficiency under single-turnover conditions in the presence of 10 mM $MgCl_2$. The reaction rates of both enzymes were approximated by single exponential functions characterized by similar rate constants (0.39 min^{-1} and 0.50 min^{-1}, for **17E(S)** and **17E(L)**, respectively) in the solution containing no organic additives. This observation indicates the absence of kinetically trapped alternate conformations of the extended DNAzyme **17E(L)**. In contrast to the hammerhead ribozyme, **17E(S)** did not show an increase in the single-turnover rate in the presence of PEG8000 (0.20 min^{-1}) and showed a decreased rate in the presence of PEG200 (0.097 min^{-1}), as shown in Figure 2b. The same effects of the presence of PEGs in solution were observed for **17E(L)** (0.39 min^{-1} with PEG8000 and 0.11 min^{-1} with PEG200).

The rate of the substrate cleavage catalyzed by **17E(S)** under multiple-turnover conditions did not show any marked change in the presence of PEG8000, and it showed a reduction in the presence of PEG200 (Figure 2c). When other mixed solutions were investigated, all mixed solutions, except those containing PEG20000 and Dex, substantially decreased the cleavage rate (Table 2). Specifically, addition of amide compounds almost completely eliminated the catalytic activity and addition of primary alcohols or aprotic compounds greatly decreased the turnover rate. The decreases in turnover rate observed in mixed solutions were characterized by the decreases in single-turnover rate (Table S1). This result suggests that the turnover rate is primarily determined by the single-turnover activity and that the effect of mixed solutions on product release is unlikely to be important. Correlation analyses revealed that the turnover rate (and also the single-turnover rate) had no significant correlation with the values of solution properties (Figure S2a–c). In addition, although the catalytic activity was observed to increase with the size of the additives, only a weak correlation was determined to exist with the additives' molecular weight (Figure S2d). It was also found that the effects of large PEGs and Dex were particularly pronounced in the presence of 1 mM $MgCl_2$, with 16-, 11-, 8.2-, and 5.1-fold rate enhancements in the presence of PEG20000, PEG8000, PEG2000, and Dex, respectively (Table 2).

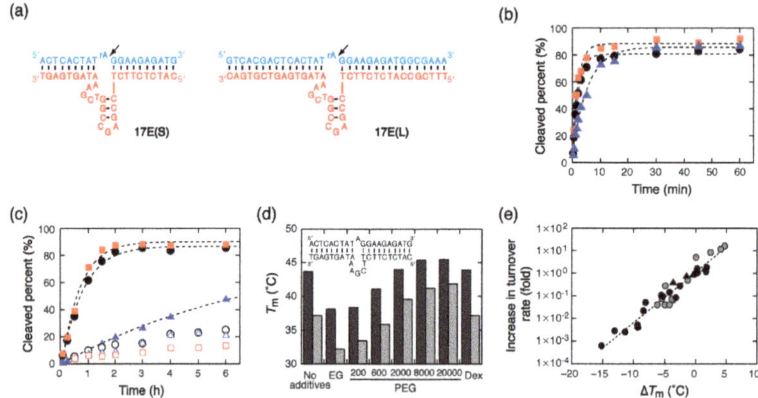

Figure 2. (**a**) Structures of the 17E DNAzymes and their substrates used in this study. The cleavage sites are indicated by arrowheads. (**b**) Kinetic traces for substrate cleavage catalyzed by **17E(L)** under single-turnover conditions in the presence of 10 mM $MgCl_2$ and in the absence (black circles) and presence of PEG8000 (red squares) or PEG200 (blue triangles). (**c**) Kinetic traces for substrate cleavage catalyzed by **17E(S)** (closed symbols) and **17E(L)** (open symbols) under multiple-turnover conditions in the presence of 10 mM $MgCl_2$ and in the absence (black circles) and presence of PEG8000 (red squares) or PEG200 (blue triangles). (**d**) T_m values of the DNA structure presented in the inset evaluated in solutions containing 10 mM $MgCl_2$ (black) or 1 mM $MgCl_2$ (gray), with and without EG, PEGs, or Dex. (**e**) Plot of the increments of the turnover rate of the **17E(S)**-catalyzed reaction against ΔT_m in the presence of 10 mM $MgCl_2$ (black) or 1 mM $MgCl_2$ (gray), excluding the data for reactions characterized by a very slow rate. The data obtained using 10% or 30% PEG solutions (PEG8000 and PEG200) and EtOH solution are also included (triangles). The correlation coefficient of a linear fit is 0.97.

Table 2. Fold increase in the rate of the **17E(S)**-catalyzed turnover reaction in mixed solutions over the rate in the absence of organic additives.

Solution	10 mM $MgCl_2$	1 mM $MgCl_2$	Solution	10 mM $MgCl_2$	1 mM $MgCl_2$	Solution	10 mM $MgCl_2$	1 mM $MgCl_2$
PEG20000	1.8	16	PDO	0.14	0.073	FA	<0.001	<0.03
PEG8000	1.2	11	MME	0.0041	<0.03	DMF	<0.001	<0.03
PEG2000	1.0	8.2	DME	0.052	0.28	AcAm	<0.001	<0.03
PEG600	0.27	0.49	MeOH	0.022	0.076	AcCN	<0.001	<0.03
PEG200	0.095	0.064	EtOH	0.0026	0.041	DMSO	0.0051	<0.03
EG	0.11	0.040	PrOH	<0.001	<0.03	DOX	0.0029	<0.03
Glyc	0.083	0.040	Urea	<0.001	<0.03	Dex	1.6	5.1

To understand the effects of the mixed solutions on the turnover rate of the **17E(S)**-catalyzed reaction, the thermal stability of the complex between the DNAzyme and its substrate was studied using a model DNA sequence of base pairs that gave the thermal melting temperature (T_m) of a two-state transition. Analysis of the melting curve indicated that the DNA structure was destabilized in the presence of small PEGs or EG, but it was slightly stabilized in the presence of large PEGs or Dex (Figure 2d). When $MgCl_2$ concentration decreased to 1 mM, the stabilization effects of large PEGs increased. These results are in agreement with previous reports according to which small PEGs reduce the stability of DNA base pairing [41–43] and large PEGs are able to increase the stability at low salt concentrations [44]. Figure 2e shows the turnover rate increments plotted against the T_m changes (ΔT_m) obtained in the mixed solutions, excluding the data for reactions characterized by a very slow rate (with a rate constant smaller than 1×10^{-3} h^{-1}) as a degree of uncertainty exists for them. The turnover rate decreased as the T_m decreased, and the correlation plots for the solutions containing 10 mM $MgCl_2$ and those containing 1 mM $MgCl_2$ overlapped with each other. Furthermore, the data obtained with different PEG8000, PEG200, or EtOH concentrations (10% and 30%) fell on the correlation line. The high level of correlation between the increments of the turnover rate and the ΔT_m of a model DNA sequence of base pairs indicates the importance of the stability of DNA interactions for the catalytic activity. In addition, the analysis using the free energy change (ΔG°) associated with complex formation inferred from the melting curve showed that the turnover rate constant is also correlated with the value of ΔG° (Figure S2e). These correlation plots may be useful for predicting the turnover rate in other mixed solutions or in solutions containing Mg^{2+} at concentrations not employed in this study. They may also be used to find the conditions that can enhance the turnover activity of the DNAzyme. Conversely, the presence of organic additives did not increase the turnover rate of the reaction catalyzed by the extended DNAzyme **17E(L)**, as shown in Figure 2c, regardless of the ability of the mixed solutions to cause changes in the single-turnover rate comparable to those of **17E(S)** (Figure S2f).

2.3. Substrate Ligation by Ribozyme and DNAzyme

Several types of RNA-ligating ribozymes exist that catalyze the formation of covalent bonds between two RNA strands. This study investigated the ligase ribozyme derived from the R3C ribozyme obtained through in vitro selection [45,46]. This ribozyme catalyzes the formation of a phosphodiester bond between the 5′-triphosphate and 3′-hydroxyl termini of substrate RNAs in the presence of Mg^{2+}. **R3C(H)** catalyzes the ligation of the ribozyme with a substrate RNA, and **R3C(T)** catalyzes the ligation of two substrate RNAs (Figure 3a). **R3C(H)** was used for the analysis of single-turnover rate, and **R3C(T)** was used to analyze the turnover rate. We also investigated **9DB1** shown in Figure 3a, derived from the 9DB1 DNAzyme obtained through in vitro selection [47]. This DNAzyme catalyzes the formation of a 3′,5′-phosphodiester bond of two substrate RNAs in the presence of Mg^{2+}, and it was designed to catalyze the ligation of the same substrates as the ligase ribozymes.

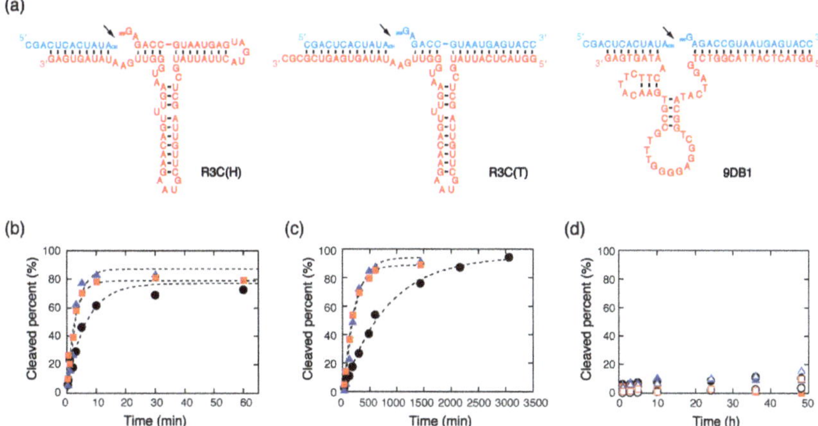

Figure 3. (a) Structures of the ligase ribozymes and DNAzyme and their substrates used in this study. The ligation sites are indicated by arrowheads. (b) Kinetic traces for substrate ligation catalyzed by **R3C(H)** under single-turnover conditions in the presence of 10 mM $MgCl_2$ and in the absence (black circles) and presence of PEG8000 (red squares) or PEG200 (blue triangles). (c) Kinetic traces for substrate ligation catalyzed by **9DB1** under single-turnover conditions in the presence of 10 mM $MgCl_2$ and in the absence (black circles) and presence of PEG8000 (red squares) or PEG200 (blue triangles). (d) Kinetic traces for substrate ligation catalyzed by **R3C(T)** (closed symbols) and **9DB1** (open symbols) under multiple-turnover conditions in the presence of 10 mM $MgCl_2$ and in the absence (black circles) and presence of PEG8000 (red squares) or PEG200 (blue triangles).

Kinetic data showed that the single-turnover rate of the reaction catalyzed by **R3C(H)** in the presence of 10 mM MgCl$_2$ was comparable to the rates obtained for the nuclease enzymes. Addition of PEGs to the solution increased the ligation rate (Figure 3b); and the same effect was observed for many organic additives, but addition of amide compounds greatly decreased the rate (Table 3). When the MgCl$_2$ concentration decreased to 1 mM, almost no reaction was observed either with or without organic additives. Similar results were observed for **9DB1**, although the increments of the ligation rate with respect to the solution containing no organic additives were, in most cases, greater than those observed for **R3C(H)** (Figure 3c and Table 3). Correlation analyses showed a less significant correlation between the single-turnover rates and the values of solution properties, the additives' molecular weight, or base-pair stability (data not shown), suggesting that the activity of these enzymes is not simply determined by a single environmental factor. On the other hand, experiments with **R3C(T)** and **9DB1** showed no turnover activity whether or not organic additives were present (Figure 3d). In the case of these ligase enzymes, therefore, mixed solutions are not useful to enhance turnover activity, but they are when it comes to the single-turnover activity.

Table 3. Fold increase in the rates of the **R3C(H)**-catalyzed and **9DB1**-catalyzed single-turnover reactions in mixed solutions in the presence of 10 mM MgCl$_2$ over the rates in the absence of organic additives.

Solution	R3C(H)	9DB1	Solution	R3C(H)	9DB1	Solution	R3C(H)	9DB1
PEG20000	1.6	4.2	PDO	1.2	2.6	FA	<0.001	<0.001
PEG8000	2.0	4.6	MME	1.9	3.3	DMF	0.26	<0.001
PEG2000	1.6	6.8	DME	1.2	1.9	AcAm	0.30	<0.001
PEG600	1.3	4.2	MeOH	1.5	4.5	AcCN	1.8	4.5
PEG200	2.0	4.1	EtOH	2.2	2.4	DMSO	1.7	2.5
EG	1.0	2.8	PrOH	1.8	3.9	DOX	1.3	3.8
Glyc	1.2	1.1	Urea	<0.001	<0.001	Dex	1.3	0.89

3. Discussion

3.1. Classification of the Mixed Solutions Based on the Effects on Nucleic Acids

In this study, we investigated the single- and multiple-turnover kinetics of nucleic acid enzymes in various mixed aqueous solutions that contain water-soluble organic cosolvents or cosolutes. These organic additives may alter the physical properties of solution, interact with nucleic acids, and introduce areas of excluded volume [20,31]. Most mixed solutions have lower dielectric constant values than pure water, and the additives, such as large PEGs, MME, DME, EtOH PrOH, and DOX, significantly lowers the dielectric constant of solution. The dielectric constant affects the strength of electrostatic interactions, including metal ion binding to nucleic acid phosphate groups. Hydrophilic additives, such as EG, small primary alcohols, and AcCN significantly reduce the water activity of solution. Water activity affects base pairing and tertiary folding accompanied by water association or release (hydration changes). In addition, preferential binding to or exclusion of additives from nucleic acid surface also affects the stability of nucleic acid interactions. The additives, such as MME, PrOH, amide compounds, AcCN, and DOX, significantly reduce the stability of base-pairing interactions. On the other hand, large PEGs and Dex occupy large volumes and increase the viscosity of solution. Viscosity affects the rate of substrate hybridization and the excluded-volume effect facilitates nucleic acid hybridization and folding into a compact structure by increasing effective concentration and restricting conformational dynamics. Accordingly, the molecular environment of mixed solutions is assumed to influence the activity of the nucleic acid enzymes through a combination of mechanisms.

The data shown in Tables 1–3 suggest the possibility to employ mixed solutions as reaction media of these enzymes. The solutions containing large PEGs increased the catalytic activity of all tested ribozymes and DNAzymes, even although these enzymes have different catalytic core sequences, tertiary interactions, roles of metal ions, and mechanism of catalysis. The solutions containing small

PEGs, ethylene glycol derivatives, small primary alcohols, and aprotic compounds increased the turnover rate of the reactions catalyzed by **HH(S)** and the single-turnover rates catalyzed by **R3C(H)** and **9DB1**, but they reduced the turnover rate catalyzed by **17E(S)**. The solutions containing amide compounds greatly reduced the catalytic activity of these enzymes, particularly at the low $MgCl_2$ concentration of 1 mM. This classification based on the chemical structure and size of additives can provide some explanation for the effects on the activity of the nucleic acid enzymes, and the correlation analyses have further determined the environmental factors causing the observed effect of the additives on **HH(S)** and **17E(S)**. Conversely, the mixed solutions did not increase the turnover rate of the extended nuclease enzymes **HH(L)** and **17E(L)** and the ligase enzymes **R3C(T)** and **9DB1**. These observations were attributed to the reduced tendency to allow the dissociation of the long reaction products, due to the high stability of base-pairing interactions.

3.2. Comparison of the Effects of Mixed Solutions on the Nucleic Acid Enzymes

3.2.1. Effects on the Hammerhead Ribozyme

The turnover rate of the **HH(S)**-catalyzed reaction increased in many mixed solutions with respect to the solution containing no organic additives. This result indicates that many kinds of organic additives are useful for enhancing the turnover activity of the hammerhead ribozyme. It has been reported that the stability of base pairing and tertiary folding is reduced in the presence of organic solvents, particularly amide compounds that are known to substantially destabilize nucleic acid structures [31,44]. However, this destabilization effect seemed not to cause a disruption of interactions necessary for the formation of the active conformation, such as a network of hydrogen bonds [48], in the presence of 10 mM $MgCl_2$. The correlation plot of Figure 1e suggests that the dielectric constant influences the turnover rate. A simple interpretation of this correlation is that the transition state is electrostatically stabilized in less-polar media. However, our previous study identified an apparent increase in Mg^{2+}-binding affinity to **HH(S)** in mixed solutions with a low dielectric constant [34,35]. The substrate cleavage catalyzed by the hammerhead ribozyme requires the involvement of several Mg^{2+} ions, and at least one Mg^{2+} ion binds to the site with highly negative electrostatic potential located near the scissile phosphate of the substrate [49]. This mechanism is consistent with the dielectric constant effect that facilitates electrostatic Mg^{2+} binding and allows greater enhancement of the turnover activity observed at the low Mg^{2+} concentration of 1 mM. In the cellular environment, the dielectric constant is relatively low [50,51] and the concentration of Mg^{2+} is as low as 1 mM or less [52]. Thus, it is speculated that the hammerhead ribozyme is suitable for use in living cells, for example to achieve intracellular mRNA cleavage and in vivo sensing.

3.2.2. Effects on the 17E DNAzyme

The activity of **17E(S)** was not enhanced in the presence of many organic additives under both single- and multiple-turnover conditions, in contrast to the case of **HH(S)**. The result shows that increasing the activity of the 17E DNAzyme is difficult to achieve using mixed aqueous solutions. Correlation analysis revealed no significant contribution of the dielectric constant to the catalytic activity of **17E(S)**. Conversely, the correlation plot shown in Figure 2e suggests that the stability of DNA interactions strongly influences the turnover rate. However, the effects of mixed solutions on the single-turnover rate of the reaction catalyzed by the extended DNAzyme **17E(L)** were similar to those measured for **17E(S)**. Thus, the destabilization effect on base pairing between the enzyme and substrate is unlikely to be an important factor. It is possible that the mixed solutions reduce the stability of DNA interactions important for formation of the catalytically active complex. Although the detailed structure and enzymatic mechanism of the 17E DNAzyme have not been elucidated, biophysical studies have shown the importance of Mg^{2+} binding and non-canonical hydrogen bonding in the catalytic core for catalysis [53–55]. It is noted that an RNA-cleaving DNAzyme exhibiting high activity in the presence of high ethanol concentration have been reported [27,28]. The active complex of this enzyme might be stable enough to overcome the destabilization effect of ethanol, unlike the case of **17E(S)**. The turnover activity of **17E(S)** was enhanced in the presence of large PEGs and Dex, which increased the stability of DNA interactions. As the large-sized additives are characterized by an excluded-volume effect, this DNAzyme might be suitable for use as a probe anchored on the surface of a sensing material, where the excluded volume of immobilized molecules is large.

3.2.3. Effects on the R3C Ribozyme and 9DB1 DNAzyme

The turnover of the reactions catalyzed by ligases is assumed to be more difficult to achieve than it is for nucleases, because ligation leads to an increase in the number of base pairs, thus strengthening the bonding between the enzyme and reaction products. As expected, no turnovers were observed in the reactions catalyzed by **R3C(T)** and **9DB1** even in mixed solutions that greatly reduced the stability of base-pairing interactions. By contrast, many mixed solutions increased the single-turnover rates. The result indicates that the active complexes of these enzymes are stable enough to overcome the destabilization effect of the organic solvents. However, addition of amide compounds greatly decreased the ligation activity, suggesting the importance of the stability of catalytic active complex in the mixed solutions for catalysis. Enhancements of the rate of the ligation catalyzed by 9DB1 DNAzyme in mixed solutions containing ethanol, methanol, isoamyl alcohol, acetone, DMSO, or DMF have been previously reported [26]. Our results exhibited that many other organic compounds also have an ability to enhance the activity of this DNAzyme, although the activity was completely abolished at the low $MgCl_2$ concentration of 1 mM. Although the detailed mechanism and structure of the R3C ribozyme have not been elucidated, the crystal structure of 9DB1 DNAzyme has been reported, and it shows the presence of tertiary interactions in the catalytic domain; however, the role of metal ions in the reaction remains unclear [56]. The mixed solutions enhanced the catalytic activities of both **R3C(T)** and **9DB1**, implying a similar mechanism by which the environmental factors influence the activity of these enzymes, whereas correlation analyses suggested that changes in the activity of these enzymes was not determined by a single environmental factor. It is mentioned that the cleavage and ligation of RNA strands are related to RNA evolution and the RNA world, where the prebiotic environment is likely to contain organic compounds. It is fascinating to speculate on the contribution of environmental factors to the evolution of functional RNAs.

4. Materials and Methods

4.1. Materials

The hammerhead ribozyme, ligase ribozyme, and substrate RNAs having a $5'$-triphosphate group were prepared by in vitro transcription using T7 RNA polymerase and purified by polyacrylamide gel electrophoresis, as described previously [34]. Other RNAs, DNAs, and a DNA–RNA chimeric strand, purified by high performance liquid chromatography, were purchased from Hokkaido System Science (Sapporo, Japan) or Fasmac (Atsugi, Japan). All reagents used for preparing buffer solutions were purchased from Wako Chemicals (Osaka, Japan) with the following exceptions: 4-(2-hydroxyethyl)-1-piperazineethanesulfonic acid (HEPES), 3-[4-(2-hydroxyethyl)-1-piperazinyl]propanesulfonic acid (HEPPS), and the disodium salt of ethylenediaminetetraacetic acid (Na_2EDTA) were purchased from Dojindo (Kumamoto, Japan); PEG with an average molecular weight of 8×10^3 was purchased from MP Biomedicals (Tokyo, Japan); 2-methoxyethanol and 1,2-dimethoxyethane were purchased from TCI (Tokyo, Japan); and dextran with an average molecular weight of 1×10^4 from Sigma-Aldrich (St. Louis, MO, USA). Values of solution properties (relative dielectric constant, water activity, and viscosity) were determined, as described previously [44].

4.2. Kinetic Studies on the Ribozymes and DNAzymes

Kinetics of the reactions catalyzed by hammerhead ribozymes and 17E DNAzymes were measured in buffer solutions comprising 50 mM HEPES, 50 mM NaCl, and 0.1 mM Na_2EDTA. The pH of this solution was adjusted to 7.0 at 37 °C. Kinetics of the reactions catalyzed by R3C ligase ribozymes and 9DB1 DNAzymes were measured in buffer solutions comprising 50 mM HEPPS, 50 mM NaCl, and 0.1 mM Na_2EDTA. The pH of this solution was adjusted to 8.0 at 37 °C. Mixed aqueous solutions were prepared by adding a cosolvent or cosolute at a concentration of 20% by weight, unless otherwise stated.

Substrate strands were fluorescently-labeled either with carboxyfluorescein (FAM) or carboxytetramethylrhodamine (TAMRA) at the 5′-end. For single-turnover reactions, substrate cleavage was measured using a 2 μM concentration of ribozyme (or DNAzyme) and a 0.1 μM concentration of substrate, and substrate ligation was measured using a 1 μM concentration of ribozyme (or DNAzyme) and a 0.1 μM concentration of substrate. For multiple-turnover reactions, substrate cleavage was measured using a 5 nM concentration of ribozyme (or DNAzyme) and a 0.1 μM concentration of substrate, and substrate ligation was measured using a 10 nM concentration of ribozyme (or DNAzyme) and a 0.1 μM concentration of substrate or a 0.1 μM concentration of fluorescently-labeled 5′-fragment substrate and a 1 μM concentration of non-labeled 3′-fragment substrate to be used for the ligation of two substrate RNAs. Before initiating the reaction by adding $MgCl_2$ at 37 °C, the enzyme and substrate(s) were annealed in the buffer solution at 60 °C, in the case of the nuclease enzymes, or 80 °C, in the case of the ligase enzymes. The reactions were quenched by mixing the reaction mixture with a 90% formamide solution containing 100 mM Na_2EDTA. The quenched solutions were then loaded onto a 20% polyacrylamide gel containing 7 M urea. After gel electrophoresis, the fluorescence emission from fluorescently-labeled strands was quantified using a fluorescence scanner (FLA-7000, Fujifilm, Tokyo, Japan).

The rate constant (k) for substrate cleavage or ligation was determined from a plot of the fraction of the reaction product against time, fitted to a single-exponential equation $f = f_0 + (f_{max} - f_0)\{1 - \exp(-kt)\}$, where f_0 is the fraction at time zero and f_{max} is the fraction at the end of the reaction. A linear approximation was used to calculate the k value of very slow reactions.

4.3. Thermal Stability Measurements

Thermal melting curve of a DNA structure was obtained by monitoring absorption at 260 nm using a spectrophotometer (UV1800, Shimadzu, Kyoto, Japan) equipped with a temperature controller. Oligonucleotides at a strand concentration of 2 μM were prepared using a buffer solution comprising 50 mM HEPES, 50 mM NaCl, 10 mM $MgCl_2$, and 0.1 mM Na_2EDTA. The pH of this solution was adjusted to 7.0. Mixed solutions were prepared at a concentration of organic additives of 20% by weight, unless otherwise stated. The solutions were placed in a cuvette sealed with an adhesive sheet. After annealing at a cooling rate of 2 °C min^{-1}, the melting curve was measured at a heating rate of 0.5 °C min^{-1}. The value of T_m, defined as the temperature at which half of the structure was denatured, and $\Delta G°$ at 37 °C were determined from the shape of the melting curve, as described previously [43].

5. Conclusions

The molecular environment of mixed aqueous solutions that contain organic compounds can enhance the catalytic activity of the hammerhead ribozyme, 17E DNAzyme, R3C ribozyme, and 9DB1 DNAzyme. In particular, the addition of large PEGs to aqueous reaction media effectively increased the rate of the reactions catalyzed by these enzymes, but systematic investigation using different kinds of mixed solutions and correlation analyses revealed that the effect of PEGs on the catalytic reaction rates had different origins with different enzymes. The activity of the hammerhead ribozyme was enhanced by a decrease in dielectric constant of the medium, because such decrease led to stronger Mg^{2+} binding interactions with the ribozyme. The activity of the 17E DNAzyme increased in the presence of large-sized additives, as these compounds caused an increase in the stability of DNA interactions. Although destabilization of base pairing between the enzyme and substrate was expected to beneficial for the catalytic turnover, the mixed solutions did not increase the turnover rate of the hammerhead ribozyme and 17E DNAzyme having extended base pairing. The ribozyme and DNAzyme ligases evaluated in this study also showed no significant turnover activity, but their single-turnover rates increased in many mixed solutions with respect to the solution containing no organic additives. In this case, however, the effects did not seem to be determined by a single environmental factor. These results suggest that the mixed solutions can be widely used to enhance the activity of nucleic acid enzymes, such as by modulating the physical properties of a solution and the stability of nucleic acid interactions. The correlation analysis data appears to be a useful tool for predicting the effect of mixed solutions, including those not evaluated in this study, on the activity of these enzymes. These findings provide insight into the catalytic activity of nucleic acid enzymes under various conditions that are applicable to the medical and technology fields, for example in living cells and in biosensors.

Supplementary Materials: The following are available online at www.mdpi.com/2073-4344/7/12/355/s1, Table S1: Increments of the single-turnover rates of the **17E(S)**-catalyzed reaction, Figure S1: Correlation plots of the turnover rates of the **HH(S)**-catalyzed reaction, Figure S2: Correlation plots of the turnover rates of the **17E(S)**-catalyzed reaction, and the increments of the single-turnover rates of the reactions catalyzed by **17E(S)** and **17E(L)**.

Acknowledgments: We thank Junpei Ueno for technical assistance. This work was supported in part by Grants-in-Aid for Scientific Research from Japan Society for the Promotion of Science (JSPS) (24550200 and 15K05575) and MEXT-Supported Program for the Strategic Research Foundation at Private Universities (2009–2014).

Author Contributions: S.N. conceived and designed the experiments; S.N., M.H. and M.K. performed the experiments; S.N., M.H. and M.K. analyzed the data; S.N. and N.S. contributed reagents and materials tools; S.N. wrote the paper.

Conflicts of Interest: The authors declare no conflict of interest.

References

1. Jäschke, A.; Seelig, B. Evolution of DNA and RNA as catalysts for chemical reactions. *Curr. Opin. Chem. Biol.* **2000**, *4*, 257–262.
2. Citti, L.; Rainaldi, G. Synthetic hammerhead ribozymes as therapeutic tools to control disease genes. *Curr. Gene Ther.* **2005**, *5*, 11–24. [CrossRef] [PubMed]
3. Fokina, A.A.; Stetsenko, D.A.; Francois, J.C. DNA enzymes as potential therapeutics: Towards clinical application of 10–23 DNAzymes. *Expert Opin. Biol. Ther.* **2015**, *15*, 689–711. [CrossRef] [PubMed]
4. Li, J.; Lu, Y. A highly sensitive and selective catalytic DNA biosensor for lead ions. *J. Am. Chem. Soc.* **2001**, *122*, 10466–10467. [CrossRef]
5. Rueda, D.; Walter, N.G. Fluorescent energy transfer readout of an aptazyme-based biosensor. *Methods Mol. Biol.* **2006**, *335*, 289–310. [PubMed]
6. Li, S.; Nosrati, M.; Kashani-Sabet, M. Knockdown of telomerase RNA using hammerhead ribozymes and RNA interference. *Methods Mol. Biol.* **2008**, *405*, 113–131.
7. Liu, J.; Cao, Z.; Lu, Y. Functional nucleic acid sensors. *Chem. Rev.* **2009**, *109*, 1948–1998. [CrossRef] [PubMed]
8. Lan, T.; Lu, Y. Metal ion-dependent DNAzymes and their applications as biosensors. *Met. Ions Life Sci.* **2012**, *10*, 217–248. [PubMed]
9. Gong, L.; Zhao, Z.; Lv, Y.F.; Huan, S.Y.; Fu, T.; Zhang, X.B.; Shen, G.L.; Yu, R.Q. DNAzyme-based biosensors and nanodevices. *Chem. Commun.* **2015**, *51*, 979–995. [CrossRef] [PubMed]
10. Liu, J.; Lu, Y. Adenosine-dependent assembly of aptazyme-functionalized gold nanoparticles and its application as a colorimetric biosensor. *Anal. Chem.* **2004**, *76*, 1627–1632. [CrossRef] [PubMed]
11. Win, M.N.; Smolke, C.D. A modular and extensible RNA-based gene-regulatory platform for engineering cellular function. *Proc. Natl. Acad. Sci. USA* **2007**, *104*, 14283–14288. [CrossRef] [PubMed]
12. Sun, C.; Liu, X.; Feng, K.; Jiang, J.; Shen, G.; Yu, R. An aptazyme-based electrochemical biosensor for the detection of adenosine. *Anal. Chim. Acta* **2010**, *669*, 87–93. [CrossRef] [PubMed]
13. Auslander, S.; Ketzer, P.; Hartig, J.S. A ligand-dependent hammerhead ribozyme switch for controlling mammalian gene expression. *Mol. Biosyst.* **2010**, *6*, 807–814. [CrossRef] [PubMed]
14. Bevilacqua, P.C.; Brown, T.S.; Nakano, S.; Yajima, R. Catalytic roles for proton transfer and protonation in ribozymes. *Biopolymers* **2004**, *73*, 90–109. [CrossRef] [PubMed]
15. Doudna, J.A.; Lorsch, J.R. Ribozyme catalysis: Not different, just worse. *Nat. Struct. Mol. Biol.* **2005**, *12*, 395–402. [CrossRef] [PubMed]
16. Lilley, D.M. Structure, folding and mechanisms of ribozymes. *Curr. Opin. Struct. Biol.* **2005**, *15*, 313–323. [CrossRef] [PubMed]

17. Wedekind, J.E. Metal ion binding and function in natural and artificial small RNA enzymes from a structural perspective. *Met. Ions Life Sci.* **2011**, *9*, 299–345. [PubMed]

18. Auffinger, P.; Grover, N.; Westhof, E. Metal ion binding to RNA. *Met. Ions Life Sci.* **2011**, *9*, 1–35. [PubMed]

19. Ward, W.L.; Plakos, K.; DeRose, V.J. Nucleic acid catalysis: Metals, nucleobases, and other cofactors. *Chem. Rev.* **2014**, *114*, 4318–4342. [CrossRef] [PubMed]

20. Nakano, S.; Miyoshi, D.; Sugimoto, N. Effects of molecular crowding on the structures, interactions, and functions of nucleic acids. *Chem. Rev.* **2014**, *114*, 2733–2758. [CrossRef] [PubMed]

21. Winship, P.R. An improved method for directly sequencing PCR amplified material using dimethyl sulphoxide. *Nucl. Acids Res.* **1989**, *17*, 1266. [CrossRef] [PubMed]

22. Jensen, M.A.; Fukushima, M.; Davis, R.W. DMSO and betaine greatly improve amplification of GC-rich constructs in *de novo* synthesis. *PLoS ONE* **2010**, *5*, e11024. [CrossRef] [PubMed]

23. Dave, N.; Liu, J. Fast molecular beacon hybridization in organic solvents with improved target specificity. *J. Phys. Chem. B* **2010**, *114*, 15694–15699. [CrossRef] [PubMed]

24. Zhang, T.; Shang, C.; Duan, R.; Hakeem, A.; Zhang, Z.; Lou, X.; Xia, F. Polar organic solvents accelerate the rate of DNA strand replacement reaction. *Analyst* **2015**, *140*, 2023–2028. [CrossRef] [PubMed]

25. Abe, H.; Abe, N.; Shibata, A.; Ito, K.; Tanaka, Y.; Ito, M.; Saneyoshi, H.; Shuto, S.; Ito, Y. Structure formation and catalytic activity of DNA dissolved in organic solvents. *Angew. Chem. Int. Ed. Engl.* **2012**, *51*, 6475–6479. [CrossRef] [PubMed]

26. Behera, A.K.; Schlund, K.J.; Mason, A.J.; Alila, K.O.; Han, M.; Grout, R.L.; Baum, D.A. Enhanced deoxyribozyme-catalyzed RNA ligation in the presence of organic cosolvents. *Biopolymers* **2013**, *99*, 382–391. [CrossRef] [PubMed]

27. Zhou, W.; Saran, R.; Chen, Q.; Ding, J.; Liu, J. A new Na^+-dependent RNA-cleaving DNAzyme with over 1000-fold rate acceleration by ethanol. *ChemBioChem* **2016**, *17*, 159–163. [CrossRef] [PubMed]

28. Yu, T.; Zhou, W.; Liu, J. Ultrasensitive DNAzyme-based Ca^{2+} detection boosted by ethanol and a solvent compatible scaffold for aptazyme design. *ChemBioChem* **2017**. [CrossRef] [PubMed]

29. Leamy, K.A.; Assmann, S.M.; Mathews, D.H.; Bevilacqua, P.C. Bridging the gap between in vitro and in vivo RNA folding. *Q. Rev. Biophys.* **2016**, *49*, e10. [CrossRef] [PubMed]

30. Nakano, S.; Sugimoto, N. Model studies of the effects of intracellular crowding on nucleic acid interactions. *Mol. Biosyst.* **2016**, *13*, 32–41. [CrossRef] [PubMed]

31. Nakano, S.; Sugimoto, N. The structural stability and catalytic activity of DNA and RNA oligonucleotides in the presence of organic solvents. *Biophys. Rev.* **2016**, *8*, 11–23. [CrossRef] [PubMed]

32. Osborne, E.M.; Schaak, J.E.; DeRose, V.J. Characterization of a native hammerhead ribozyme derived from *Schistosomes*. *RNA* **2005**, *11*, 187–196. [CrossRef] [PubMed]

33. Nelson, J.A.; Uhlenbeck, O.C. Minimal and extended hammerheads utilize a similar dynamic reaction mechanism for catalysis. *RNA* **2008**, *14*, 43–54. [CrossRef] [PubMed]

34. Nakano, S.; Karimata, H.T.; Kitagawa, Y.; Sugimoto, N. Facilitation of RNA enzyme activity in the molecular crowding media of cosolutes. *J. Am. Chem. Soc.* **2009**, *131*, 16881–16888. [CrossRef] [PubMed]

35. Nakano, S.; Kitagawa, Y.; Yamashita, H.; Miyoshi, D.; Sugimoto, N. Effects of cosolvents on the folding and catalytic activities of the hammerhead ribozyme. *ChemBioChem* **2015**, *16*, 1803–1810. [CrossRef] [PubMed]

36. Santoro, S.W.; Joyce, G.F. A general purpose RNA-cleaving DNA enzyme. *Proc. Natl. Acad. Sci. USA* **1997**, *94*, 4262–4266. [CrossRef] [PubMed]

37. Li, J.; Zheng, W.; Kwon, A.H.; Lu, Y. In vitro selection and characterization of a highly efficient Zn(II)-dependent RNA-cleaving deoxyribozyme. *Nucl. Acids Res.* **2000**, *28*, 481–488. [CrossRef] [PubMed]

38. Hwang, K.; Wu, P.; Kim, T.; Lei, L.; Tian, S.; Wang, Y.; Lu, Y. Photocaged DNAzymes as a general method for sensing metal ions in living cells. *Angew. Chem. Int. Ed. Engl.* **2014**, *53*, 13798–13802. [CrossRef] [PubMed]

39. Zhou, W.; Chen, Q.; Huang, P.J.; Ding, J.; Liu, J. DNAzyme hybridization, cleavage, degradation, and sensing in undiluted human blood serum. *Anal. Chem.* **2015**, *87*, 4001–4007. [CrossRef] [PubMed]

40. Wang, X.Y.; Niu, C.G.; Guo, L.J.; Hu, L.Y.; Wu, S.Q.; Zeng, G.M.; Li, F. A fluorescence sensor for lead(II) ions determination based on label-free gold nanoparticles (GNPs)-DNAzyme using time-gated mode in aqueous solution. *J. Fluoresc.* **2017**, *27*, 643–649. [CrossRef] [PubMed]

41. Nakano, S.; Karimata, H.; Ohmichi, T.; Kawakami, J.; Sugimoto, N. The effect of molecular crowding with nucleotide length and cosolute structure on DNA duplex stability. *J. Am. Chem. Soc.* **2004**, *126*, 14330–14331. [CrossRef] [PubMed]

42. Knowles, D.B.; LaCroix, A.S.; Deines, N.F.; Shkel, I.; Record, M.T., Jr. Separation of preferential interaction and excluded volume effects on DNA duplex and hairpin stability. *Proc. Natl. Acad. Sci. USA* **2011**, *108*, 12699–12704. [CrossRef] [PubMed]

43. Nakano, S.; Yamaguchi, D.; Tateishi-Karimata, H.; Miyoshi, D.; Sugimoto, N. Hydration changes upon DNA folding studied by osmotic stress experiments. *Biophys. J.* **2012**, *102*, 2808–2817. [CrossRef] [PubMed]

44. Nakano, S.; Kitagawa, Y.; Miyoshi, D.; Sugimoto, N. Hammerhead ribozyme activity and oligonucleotide duplex stability in mixed solutions of water and organic compounds. *FEBS Open Bio* **2014**, *4*, 643–650. [CrossRef] [PubMed]

45. Rogers, J.; Joyce, G.F. The effect of cytidine on the structure and function of an RNA ligase ribozyme. *RNA* **2001**, *7*, 395–404. [CrossRef] [PubMed]

46. Kurihara, E.; Uchida, S.; Umehara, T.; Tamura, K. Development of a functionally minimized mutant of the R3C ligase ribozyme offers insight into the plausibility of the RNA world hypothesis. *Biology* **2014**, *3*, 452–465. [CrossRef] [PubMed]

47. Purtha, W.E.; Coppins, R.L.; Smalley, M.K.; Silverman, S.K. General deoxyribozyme-catalyzed synthesis of native 3′-5′ RNA linkages. *J. Am. Chem. Soc.* **2005**, *127*, 13124–13125. [CrossRef] [PubMed]

48. Bevers, S.; Xiang, G.; McLaughlin, L.W. Importance of specific adenosine $N3$-nitrogens for efficient cleavage by a hammerhead ribozyme. *Biochemistry* **1996**, *35*, 6483–6490. [CrossRef] [PubMed]

49. Lee, T.S.; Lopez, C.S.; Giambasu, G.M.; Martick, M.; Scott, W.G.; York, D.M. Role of Mg^{2+} in hammerhead ribozyme catalysis from molecular simulation. *J. Am. Chem. Soc.* **2008**, *130*, 3053–3064. [CrossRef] [PubMed]

50. Asami, K.; Hanai, T.; Koizumi, N. Dielectric properties of yeast cells. *J. Membr. Biol.* **1976**, *28*, 169–180. [CrossRef] [PubMed]

51. Cuervo, A.; Dans, P.D.; Carrascosa, J.L.; Orozco, M.; Gomila, G.; Fumagalli, L. Direct measurement of the dielectric polarization properties of DNA. *Proc. Natl. Acad. Sci. USA* **2014**, *111*, E3624–E3630. [CrossRef] [PubMed]

52. Grubbs, R.D. Intracellular magnesium and magnesium buffering. *Biometals* **2002**, *15*, 251–259. [CrossRef] [PubMed]

53. Raz, M.H.; Hollenstein, M. Probing the effect of minor groove interactions on the catalytic efficiency of DNAzymes 8–17 and 10–23. *Mol. Biosyst.* **2015**, *11*, 1454–1461. [CrossRef] [PubMed]

54. Kim, H.K.; Rasnik, I.; Liu, J.; Ha, T.; Lu, Y. Dissecting metal ion-dependent folding and catalysis of a single DNAzyme. *Nat. Chem. Biol.* **2007**, *3*, 763–768. [CrossRef] [PubMed]

55. Nakano, S.; Watabe, T.; Sugimoto, N. Modulation of the ribozyme and deoxyribozyme activities using tetraalkylammonium ions. *ChemPhysChem* **2017**. [CrossRef] [PubMed]

56. Ponce-Salvatierra, A.; Wawrzyniak-Turek, K.; Steuerwald, U.; Hobartner, C.; Pena, V. Crystal structure of a DNA catalyst. *Nature* **2016**, *529*, 231–234. [CrossRef] [PubMed]

 *catalysts*

Article

Effect of Iminodiacetic Acid-Modified Nieuwland Catalyst on the Acetylene Dimerization Reaction

Yanhe You, Juan Luo, Jianwei Xie * and Bin Dai *

Key Laboratory for Green Processing of Chemical Engineering of Xinjiang Bingtuan,
School of Chemistry and Chemical Engineering, Shihezi University, North 4th road,
Shihezi 832003, China; shzu_yanheyou@sina.com (Y.Y.); shzu_juanluo@sina.com (J.L.)
* Correspondence: cesxjw@foxmail.com (J.X.); db_tea@shzu.edu.cn (B.D.);
  Tel.: +86-993-2057213 (J.X. & B.D.); Fax: +86-993-2057270 (J.X. & B.D.)

Received: 19 November 2017; Accepted: 16 December 2017; Published: 19 December 2017

Abstract: The iminodiacetic acid-modified Nieuwland catalyst not only improves the conversion of acetylene but also increases the selectivity of monovinylacetylene (MVA). A catalyst system containing 4.5% iminodiacetic acid exhibited excellent performance, and the yield of MVA was maintained at 32% after 24 h, producing an increase in the yield by 12% relative to the Nieuwland catalyst system. Based on a variety of characterization methods analysis of the crystal precipitated from the catalyst solution, it can be inferred that the outstanding performance and lifetime of the catalyst system was due to the presence of iminodiacetic acid, which increases the electron density of Cu^+ and adjusts the acidity of the catalytic solution.

Keywords: acetylene dimerization; selectivity; performance; monovinylacetylene; iminodiacetic acid-modified Nieuwland catalyst

---

1. Introduction

As one of the seven main synthetic rubber materials, chloroprene (CR) possesses good mechanical properties and chemical stability [1] and is widely used in the adhesive and automobile industries and also in other fields [2,3]. Acetylene dimerization facilitated by the Nieuwland catalyst to produce monovinylacetylene (MVA) is the key step in the acetylene-based process for CR synthesis [4]. The main process of this reaction is shown in the Scheme 1. CuCl, NH_4Cl (or KCl) and water constitute the traditional Nieuwland catalyst, which has several advantages, such as being environmentally friendly, convenient to prepare, and low reaction temperature [5,6]. However, the drawbacks of this catalytic system include low acetylene conversion rates and low MVA selectivity. Thus, much effort is still needed to develop a more stable and efficient Nieuwland catalytic system to further improve the activity and selectivity of the acetylene dimerization reaction.

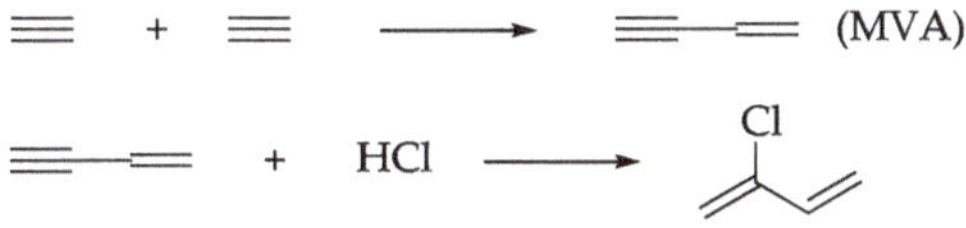

Scheme 1. Acetylene-based process for 2-chloro-1,3-butadiene production.

Because of the practical importance and the long history of the Nieuwland catalyst, recent studies on the reaction mechanism [1,6–9], catalyst structure [10,11], and factors affecting the catalytic performance [11–19] have attracted significant attention. In general, it catalyzes a homogenous transition metal-catalyzed organic transformation; the addition of a second metal and ligand/additive results in changes in the electronic structure of the active metal, thus causing change in the performance of the catalyst; the change in the electronic structure of the active metal may be due to an electron transfer between the active metal and second metal or ligand/additive. In the catalytic recycling of acetylene dimerization, the electron transfer from Cu in $Cu_nCl_{n+1}^-$ (the active component in the Nieuwland catalyst) to the π^* orbital of the C≡C bond is an important step, which acts as an activating C≡C bond. Additionally, when the electron density of Cu(I) is higher, a higher catalytic activity is obtained. Consequently, several groups have successfully improved the acetylene dimerization reaction using this strategy. For example, Tao et al. reported that the addition of urea [20], phosphine ligands [21], $LaCl_3$ [22,23], and $CeCl_3$ [24] to a Nieuwland catalyst solution can effectively enhance the acetylene conversion or MVA selectivity. Han et al. reported that the addition of DL-alanine to the Nieuwland catalyst can reduce the activity of acetylene dimerization and improve the MVA selectivity [25]. We also found that the addition of $SrCl_2$ [26], $ZnCl_2$ [27], PEGs [28], and $CuCl_2$ [9] can improve the selectivity of MVA or the lifetime of the catalyst. However, these modified catalysts can only improve either the acetylene conversion or MVA selectivity, and simultaneous improvements in conversion and selectivity have not been reported in the literature.

In this paper, we report iminodiacetic acid as an efficient ligand, which can simultaneously improve acetylene conversion and MVA selectivity, for acetylene dimerization catalyzed using the Nieuwland catalyst. With the iminodiacetic acid-modified Nieuwland catalyst, a 38.0% yield for acetylene conversion and 84.2% yield for MVA selectivity were obtained; the yields are comparatively higher than that obtained with the traditional Nieuwland catalyst. Moreover, the structures and reaction mechanism of the catalysts are also discussed. These results provided a new idea for improving the performance of Nieuwland catalyst, and this method is viable for use in both laboratory research and large industrial scales for acetylene-based CR production in the future.

2. Results and Discussion

2.1. Catalytic Activities of NC and L-NC

Initially, the catalytic performances of NC and L-NC were tested under fixed reaction conditions (space velocity of acetylene = 105 h^{-1}, reaction temperature = 80 °C, L: CuCl = 0.03:1), and the results are shown in Figure 1 and Table 1. For acetylene conversion, all five ligands promoted the initial catalytic activity of NC and enhanced the catalyst stability for acetylene dimerization. The order of the catalyst activity is L_4-NC > L_1-NC > L_3-NC > L_2-NC > L_5-NC ≈ NC. For MVA selectivity, all the L-NCs exhibited good selectivity in the range of 76–87%, whereas there was only 76% MVA selectivity for NC. This was surprising given the previous findings that a higher acetylene conversion resulted in lower MVA selectivity. Taken together, L_1 is the most efficient ligand in this reaction with 29.3% acetylene conversion and 84.6% MVA selectivity. Subsequently, the results of the effect of the amount of L_1 showed that an optimal catalytic performance for acetylene dimerization was achieved when 4.5% of L_1 was used, which is shown in Figure 2 and Table 2. The acetylene conversion and MVA selectivity were 38% and 84.2%, respectively, with a 12% increase in yield in comparison with the NC system.

Table 1. Effect of L_1–L_5 (3%) for acetylene dimerization.

Catalysts	Acetylene Conversion (%)	MVA Selectivity (%)	Yield (%)
NC	26.4	76.4	20.1
L_1-NC	29.3	84.6	24.8
L_2-NC	27.3	87.5	23.8
L_3-NC	28.8	79.4	22.9
L_4-NC	29.7	78.5	23.3
L_5-NC	26.3	87.1	22.9

Note: Space velocity of acetylene was 105 h^{-1} and reaction temperature was 80 °C. The data is the average of three independent runs.

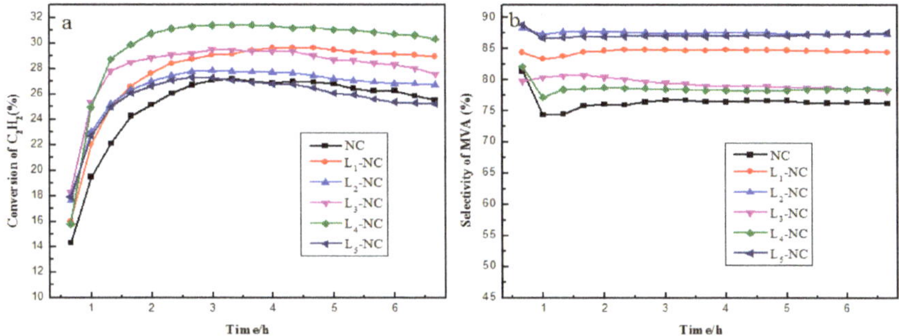

Figure 1. (**a**) Conversion of C_2H_2 and (**b**) selectivity to monovinylacetylene (MVA) in acetylene dimerization over NC and **L**-NC. (Reaction conditions: Space velocity of acetylene 105 h^{-1}, reaction temperature 80 °C, 3% of **L$_1$**–**L$_5$**).

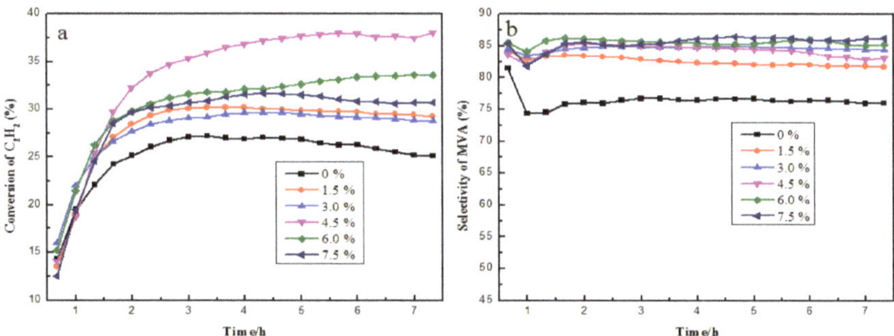

Figure 2. (**a**) Conversion of C_2H_2 and (**b**) selectivity to MVA in acetylene dimerization over 0–7.5% of **L$_1$**. (Reaction conditions: Space velocity of acetylene 105 h^{-1}, reaction temperature 80 °C).

Table 2. Effect of the amount of **L$_1$** for acetylene dimerization.

L$_1$ (%)	Acetylene Conversion (%)	MVA Selectivity (%)	Yield (%)
0	26.4	76.4	20.1
1.5	29.7	82.3	24.4
3.0	29.2	84.7	24.7
4.5	38.0	84.2	32.0
6.0	32.5	85.5	27.8
7.5	31.0	86.0	26.7

Note: Space velocity of acetylene was 105 h^{-1} and reaction temperature was 80 °C. The data is average of independent three runs.

2.2. Stability Tests

Long-term stability experiments were conducted to compare the catalytic stabilities and performances of NC, **L$_1$**-NC, and HCl-NC (Figure 3 and Table 3). **L$_1$**-NC exhibited an excellent catalytic performance and stability, and the acetylene conversion and MVA selectivity remained stable with no obvious decline observed within 24 h. In contrast, NC and HCl–NC exhibited distinctly lower catalytic performances and stabilities.

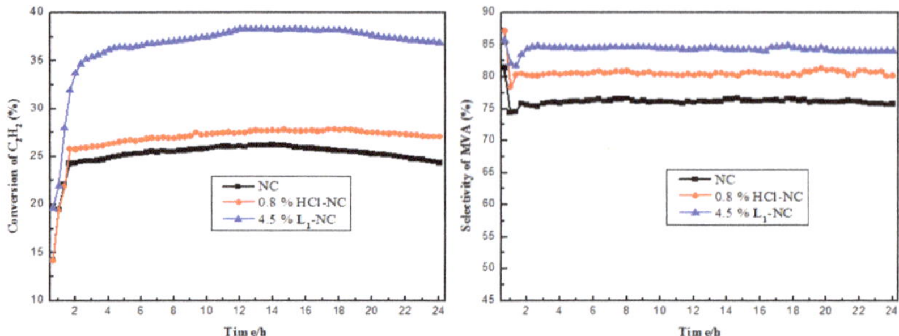

Figure 3. Catalytic performance of NC, 4.5% **L$_1$**-NC and 0.8% HCl–NC. (Reaction conditions: Space velocity of acetylene 105 h^{-1}, reaction temperature 80 °C).

Table 3. Lifetime testing for three catalysts of NC, 4.5% **L$_1$**-NC and 0.8% HCl–NC.

Catalysts	Acetylene Conversion (%)	MVA Selectivity (%)	Yield (%)
NC	25.5	76.1	19.4
HCl-NC	27.2	80.5	21.9
L$_1$-NC	37.5	84.4	31.7

Note: Space velocity of acetylene was 105 h^{-1} and reaction temperature was 80 °C. The data is average of independent three runs.

2.3. Structures of Crystals A and B

Considering the high efficiency of **L$_1$**, we investigated the relationship between the structure of NC/**L$_1$**-NC and the catalytic activity. Then, we separated crystals A and B from solutions of NC and **L$_1$**-NC via a freezing crystallization method, and we identified the composition of the crystals using FT-IR, TG, TPD-MS, XRD, XRF, and XPS analysis.

The FT-IR spectrum of crystal A is shown in Figure 4a. The peaks at 3169 cm^{-1} and 1393 cm^{-1} are characteristic peaks of O–H, and peaks representing N–H appeared at 3445 cm^{-1} and 1643 cm^{-1} [18]. In the catalyst solution, NH_4^+ was hydrolyzed to produce NH_3 ($NH_4^+ + 2H_2O = NH_3 \cdot H_2O + H_3^+O$); thus, we concluded that the crystal A contains H_2O and NH_3. As shown in Figure 4b, compared with **L$_1$**, the N–H stretching absorption peak blue-shifted by approximately 70 cm^{-1}, which indicated that N–H was likely coordinated with Cu(I) in the copper complex. The peaks at 3161 cm^{-1}, 1400 cm^{-1}, and 904 cm^{-1} were identified as the characteristic peaks of O–H in the carboxyl group. The peak at 1718 cm^{-1} is due to the C=O stretching vibration. The band located at 1064 cm^{-1} can be assigned to the C–N stretching vibration. Therefore, crystal B may contain H_2O, NH_3, and **L$_1$**. These results were also confirmed by the TPD-MS spectrum (Figure 5) because a peak for the $m/z = 17$ curve in the temperature range of 200 °C–350 °C was obtained, which suggested that the weight loss corresponded to NH_3.

To further determine the composition of the catalyst, we performed thermal analysis (TG) on crystals A and B (Figure 6). There were three weight losses for crystal A during the heating process. The first weight loss began at 100 °C, which was mainly due to the loss of crystal water [10]. Obviously, the second weight loss was due to the loss of NH_3 [18]. We calcined crystal A and crystal B in a tube furnace with nitrogen for 1 h at 450 °C; XRD studies of the residues showed that they contained CuCl and $CuCl_2$ (Figure 7). According to the TGA study for crystal A, three parts of the lost mass percentages were 0.28%, 30.04%, and 69.68%. However, there was one additional part of the weight loss between 375 °C and 405 °C for crystal B, and for crystal B, the four parts of the lost mass percentages were 1.42%, 30.40%, 3.58%, and 64.6%. XRF examination of crystals A and B showed that their Cu/Cl atomic ratios were both 2:3. Consequently, we can conclude that the compositions of crystal A and B are $Cu_2Cl_3 \cdot 6NH_3 \cdot 1/20H_2O$ and $Cu_2Cl_3 \cdot 1/10C_4H_7NO_4 \cdot 13/2NH_3 \cdot 3/10H_2O$, respectively.

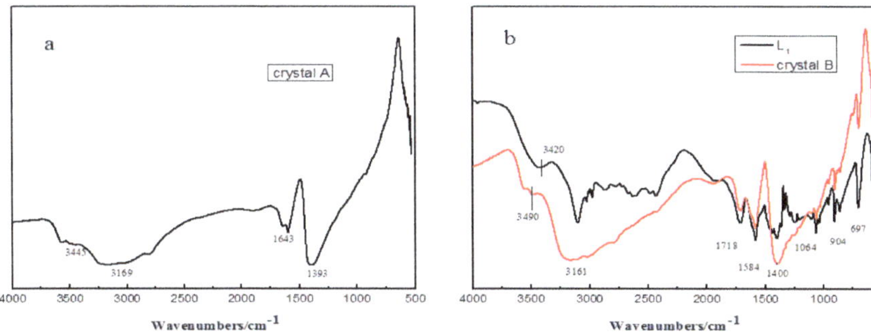

Figure 4. IR spectra of the crystal A (**a**), crystal B and **L₁** (**b**).

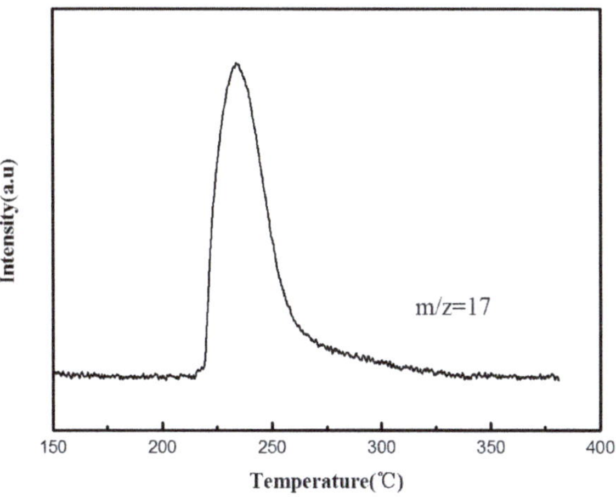

Figure 5. TPD-MS spectra of desorption of crystal A and B.

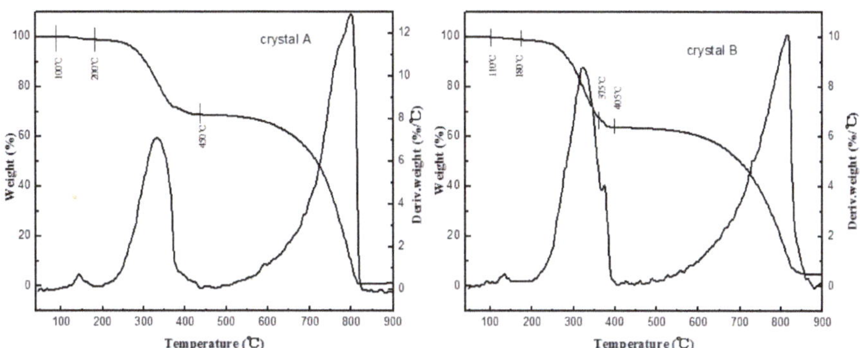

Figure 6. TG/DTG thermograms for the crystal A and crystal B.

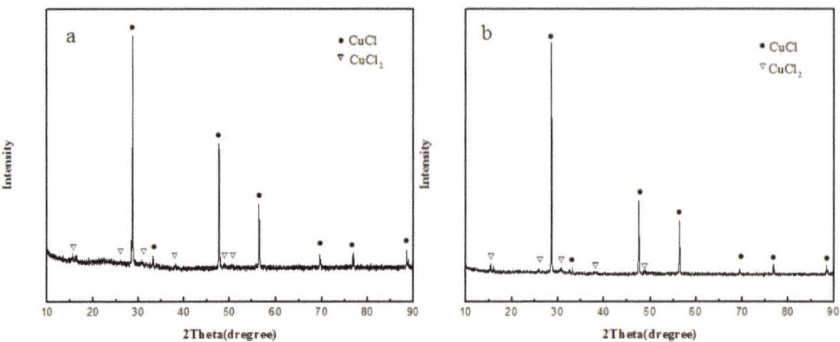

Figure 7. XRD pattern of heating product of the crystals A (**a**), B (**b**).

2.4. Action of the Ligand

As shown in Figure 8, the XPS results revealed that the binding energy of Cu $2p_{3/2}$ for crystal B exhibited a 0.26 eV higher negative shift compared with crystal A. This negative shift is attributed to the interaction between L_1 and copper ions, which increases the electron density of Cu^+ due to the transfer of electrons from L_1 to the Cu^+ center. This result was also supported by the TG analysis results of crystal B, whereas a higher desorption temperature for L_1 in crystal B was required.

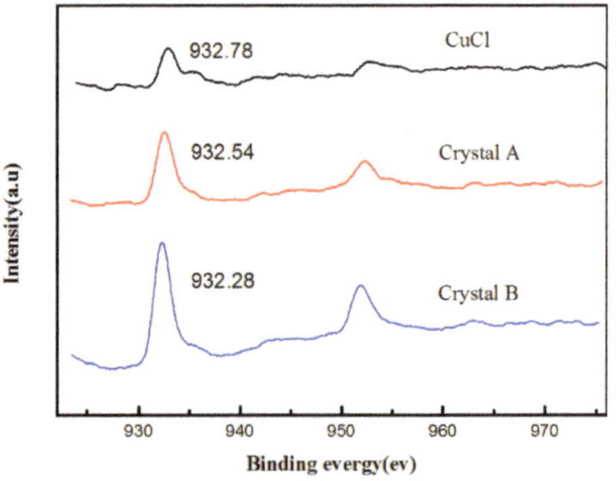

Figure 8. XPS spectra of Crystals A, B and CuCl for Cu 2p.

As mentioned above, the higher electron density of Cu(I) lead to a higher catalytic performance in acetylene dimerization. Therefore, L_1-NC exhibited the best catalytic performance (higher acetylene conversion and MVA selectivity) and stability as described in Figures 1 and 3, which was due to the strongest electronic donating and coordination capability of L_1 with Cu(I) among the five tested ligands [11]. Recently, Han et al. reported that the addition of a certain amount of hydrochloric acid to the Nieuwland catalyst can inhibit the formation of the polymer and improve the catalyst lifetime and acetylene conversion [18]. Fukuzumi et al. also reported that the addition of 2.86 mol % diethylenetriaminepentaacetic acid (DTPA) can increase the maximum yield of MVA and improve the product ratio of MVA to DVA, which may be due to the coordination of DTPA to the copper(I) active species affected the stability of the copper(I)-MVA and copper(I)-DVA complexes [1]. However, the present used additive L_1, which has a similar structure to DTPA but with more weak chelating ability and acidity, demonstrated higher catalytic performance. Therefore, both appropriate coordinate ability and acidity are important factors for an additive to enhance the catalytic activity and catalyst lifetime. At this point, L_1 is more suitable than DTPA for acetylene dimerization in large-scale application in industry.

2.5. Relationship between the Valence of Copper and Catalytic Activity

The Cu^+ ion is generally considered to be the main active component in the NC system. However, our previous studies found that Cu^{2+} also plays an important role in the acetylene dimerization reaction, and the deactivation of the anhydrous NC was caused by the transformation of Cu^+ to Cu^{2+} [9]. Hence, valence changes for copper species in fresh and used NC and L_1-NC were evaluated via high-resolution XPS, and the results are shown in Figure 9 and Table 4.

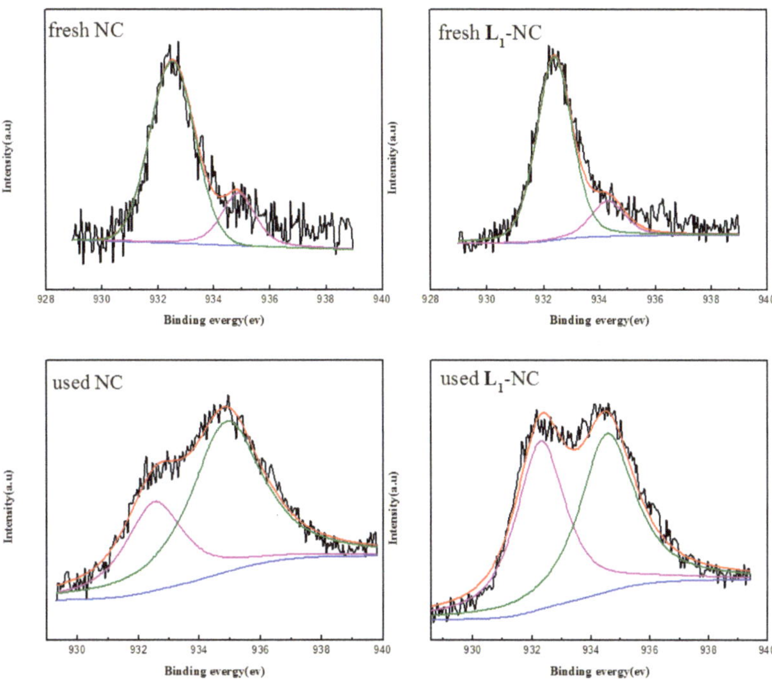

Figure 9. High-resolution XPS spectra of fresh NC, L_1-NC and used NC, L_1-NC.

Table 4. Relative content and binding energy of Cu^+ and Cu^{2+} in fresh catalyst.

Catalyst	Area%, Binding Energy (eV)			
	Cu^+		Cu^{2+}	
	Fresh	Used	Fresh	Used
NC	79.09 (932.53)	32.85 (932.51)	20.91 (934.87)	67.15 (934.88)
L_1-NC	83.14 (932.31)	48.50 (932.30)	16.86 (934.51)	51.50 (934.56)

In fresh NC, peaks at 932.53 and 934.87 eV represented Cu^+ and Cu^{2+}, which had relative contents of 79.09% and 20.91%, respectively, whereas the peak positions of Cu^+ and Cu^{2+} were at 932.31 and 934.51 eV, respectively, with relative contents of 83.14% and 16.86% for fresh L_1-NC. However, the contents of Cu^+ in the used NC and L_1-NC were 32.85% and 48.50%, respectively. The higher content of Cu^+ in used L_1-NC in comparison with that in the NC may be because L_1 inhibited the reduction of Cu^{2+} to Cu^+. These results explain why L_1-NC possesses a higher initial catalytic activity and has a longer long-term lifetime than NC.

3. Experimental

3.1. Materials and Catalyst Preparation

Reagent-grade CuCl (99%), NH_4Cl (purity $\geq$ 99.5%), iminodiacetic acid (L_1), 2-(1*H*-pyrazol-1-yl)acetic acid (L_2), 1*H*-tetrazole-1-acetic acid (L_3), diglycolic acid (L_4), and guanidineacetic acid (L_5) were purchased from Adamas (Shanghai, China).

The traditional Nieuwland catalyst was labeled as NC. The Nieuwland catalysts containing L_1, L_2, L_3, L_4, and L_5 were denoted as L_1-NC, L_2-NC, L_3-NC, L_4-NC, and L_5-NC, respectively. The initial amounts of L_1–L_5 are 3.0 mol % based on cuprous chloride (0.60 g, 0.56 g, 0.57 g, 0.60 g and 0.53 g, respectively). The structure of the five ligands was shown in Figure 10.

L_1: iminodiacetic acid

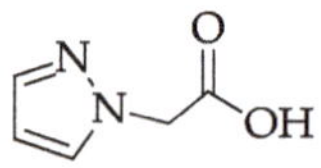

L_2: 2-(1*H*-pyrazol-1-yl)acetic acid

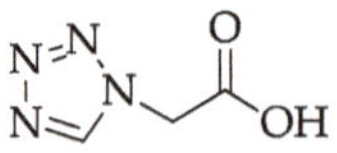

L_3: 1*H*-tetrazole-1-acetic acid

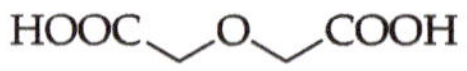

L_4: diglycolic acid

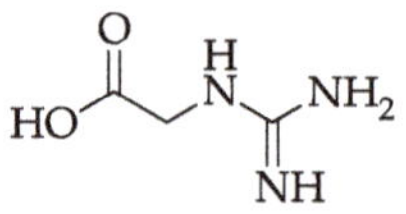

L_5: guanidineacetic acid

Figure 10. The structure of the five ligands.

3.2. The Structure of the Reactor

The reaction was performed in a self-designed glass reactor with a length of 400 mm, an outer diameter of 40 mm, an inner diameter of 20 mm. The reactor inner diameter contains the catalyst solution. A sand core baffle was used at the lower end of the reactor to increase the contact area between acetylene and the catalyst solution. The outer diameter was filled with circulating water to control the temperature of the reactor.

3.3. Catalyst Preparation and the Dimerization Reaction of Acetylene

The flow diagram of acetylene dimerization device is shown in Figure 11. Before the start of the reaction, the air in the pipe was removed by nitrogen during 30 min of continuous nitrogen flow. Then, 8 g of NH_4Cl and 15 mL of deionized water were added to the bubbled bed reactor at 80 °C under nitrogen atmosphere. After the mixture bubbled for approximately 10 min, 14.81 g of CuCl and a quantity of L_1–L_5 was added to the reactor under nitrogen atmosphere for at least 15 min until these solids got completely dissolved; then, the catalyst was obtained. The flow of C_2H_2 was regulated by a mass flowmeter. The gases in C_2H_2, such as H_2S, PH_3, and O_2, were destroyed by $K_2Cr_2O_7$ and $Na_2S_2O_4$ solutions. The acid gas and water were eliminated using a NaOH solution and a drying tube, respectively. Then, the purified C_2H_2 flowed into a preheated catalyst-containing glass reaction vessel. The gas mixture at the outlet of the reactor was analyzed via chromatography using a Shimadzu (Suzhou, China) GC-2014C equipped with a GDX-301 chromatography column and a flame ionization detector.

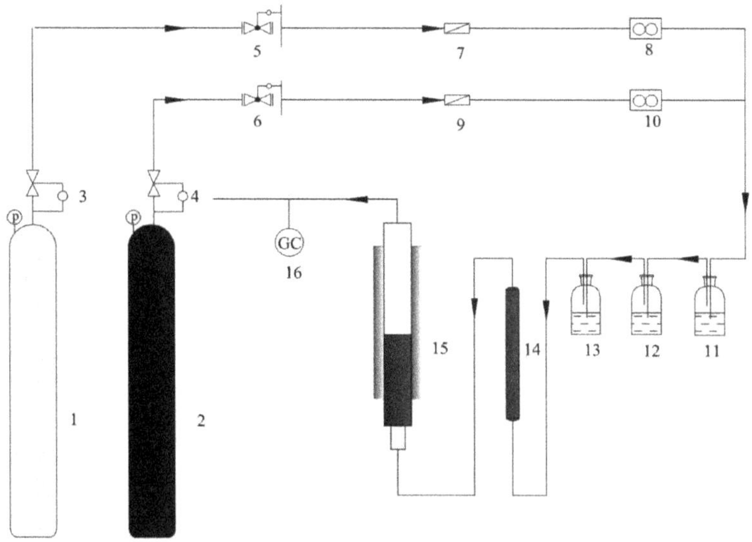

Figure 11. Schematic diagram of the C_2H_2 dimerization system: **1** C_2H_2; **2** N_2; **3, 4** partial pressure valve; **5, 6** check valve; **7, 9** gas filter; **8, 10** mass flow controller; **11** $K_2Cr_2O_7$ solution; **12** $Na_2S_2O_4$ solution; **13** NaOH solution; **14** drying tube; **15** reactor; **16** Gas Chromatograph (GC-2014C).

3.4. Analytical Methods

The conversion of C_2H_2 (X) and the selectivity to MVA (S) as the criteria of catalytic performance were defined as following equations, respectively.

$$X = [(\lambda_2 + 2\lambda_3 + 2\lambda_4 + 3\lambda_5)/(\lambda_1 + \lambda_2 + 2\lambda_3 + 2\lambda_4 + 3\lambda_5)] \times 100\%$$

$$S = [2\lambda_3/(\lambda_2 + 2\lambda_3 + 2\lambda_4 + 3\lambda_5)] \times 100\%$$

where, λ_1, λ_2, λ_3, λ_4 and λ_5 are considered as the volume fraction of acetylene, acetaldehyde, MVA, 2-chloro-1,3-butadiene and 1,5-hexadien-3-yne (DVA) in the gas product.

3.5. Separation of Crystals A and B from the NC and L_1-NC Solutions

Crystal A was obtained by gradually cooling the NC at 3 °C. It was filtrated and was washed thrice with deionized water, and then, it was dried in a vacuum oven at 55 °C. The L_1-NC was treated using the same method to obtain crystal B.

3.6. Treatment of the Used Catalyst Solutions

After the acetylene dimerization reaction occurred, the polymer in the catalyst solution containing NC or L_1-NC was filtered, and the catalyst solution was dried in a vacuum oven at 55 °C.

3.7. Catalyst Characterization

FT-IR spectra of the samples in the range of 500 to 4000 cm^{-1} were obtained using a Bruker (Karlsruhe, Gemany) Vertex70.

TG-DTG of the samples was carried out using a NETZSCH (Bavaria, Germany) STA 449 F3 Jupiter under a nitrogen atmosphere. The temperature was increased from 50 °C to 900 °C at a rate of 10 °C min^{-1}.

X-ray diffraction (XRD) patterns were collected using a Bruker (Karlsruhe, Germany) D8 Advanced X-ray diffractometer with Cu–Kα irradiation (λ = 1.5406 Å) at 40 kV and 40 mA at wide angles (10°–90° in 2θ).

A Kratos AXIS Ultra DLD spectrometer (Manchester, UK) with a monochromatized, Al–Kα X-ray source (225 W) was used to record the XPS data.

Elemental analysis of the crystals was carried out via X-ray fluorescence (XRF) using a Shimadzu (Tokyo, Japan) XRF-1800 system (Rh target) at 40 kV and 1 mA with HS Easy software.

TPD-MS experiments were carried out using a Micromeritic (Atlanta, GA, USA) ASAP 2720 apparatus equipped with a TCD detector.

4. Conclusions

We demonstrated an iminodiacetic acid-assisted aqueous Nieuwland catalyst for the acetylene dimerization reaction that can achieve a 38.0% yield for acetylene conversion and an 85% yield for MVA selectivity with an increase in yield of 12% relative to the Nieuwland catalyst system. In this transformation, iminodiacetic acid not only provides proper acidity of the catalyst solution but also prevents Cu^+ oxidation to Cu^{2+}, which are beneficial for the catalytic activity and stability. Further investigations on acetylene dimerization using this strategy are ongoing in our laboratory and will be reported in the future.

Acknowledgments: We acknowledge National Natural Science Foundation of China (No. 21463021 and 21776179) for their financial support.

Author Contributions: Jianwei Xie and Bin Dai received the project. Yanhe You and Juan Luo performed the experiments and analyzed the spectra data. Yanhe You, Jianwei Xie and Bin Dai wrote the paper.

Conflicts of Interest: The authors declare no conflict of interest.

References

1. Tachiyama, T.; Yoshida, M.; Aoyagia, T.; Fukuzumi, S. Mechanistic study on dimerization of acetylene with a Nieuwland catalyst. *Appl. Organomet. Chem.* **2008**, *22*, 205–210. [CrossRef]

2. Rattanasom, N.; Kueseng, P.; Deeprasertkul, C. Improvement of the mechanical and thermal properties of silica-filled polychloroprene vulcanizates prepared from latex system. *Appl. Polym. Sci.* **2012**, *124*, 2657–2668. [CrossRef]

3. Ismail, H.; Leong, H.C. Curing characteristics and mechanical properties of natural rubber/chloroprene rubber and epoxidized natural rubber/chloroprene rubber blends. *Polym. Test.* **2001**, *20*, 509–516. [CrossRef]

4. Arutyunyan, R.M.; Pogosyan, V.S.; Simonyan, E.H.; Atoyants, A.L.; Djigardjian, E.M. In situ monitoring of the ambient air around the chloroprene rubber industrial plant using the Tradescantia-stamen-hair mutation assay. *Mutat. Res.* **1999**, *426*, 117–120. [CrossRef]

5. Liu, J.G.; Han, M.H.; Zuo, Y.Z.; Wang, Z.W. Research progress of dimerization of acetylene to monovinylacetylene. *Chem. Ind. Eng. Prog.* **2011**, *30*, 942–947. [CrossRef]

6. Nishiwaki, K.; Kobayashi, M.; Takeuchi, T.; Matuoto, K.; Osakada, K. Nieuwland catalysts: Investigation of structure in the solid state and in solution and performance in the dimerization of acetylene. *J. Mol. Catal. A Chem.* **2001**, *175*, 73–81. [CrossRef]

7. Tokita, Y.; Okamoto, A.; Nishiwaki, K.; Kobayashi, M.; Nakamura, E. Kinetics of copper(I)-catalyzed dimerization and hydration of acetylene in water. *Bull. Chem. Soc. Jpn.* **2004**, *77*, 1395–1399. [CrossRef]

8. Tachiyama, T.; Yoshida, M.; Aoyagi, T.; Fukuzumi, S. Deuterium kinetic isotope effects and H/D exchange in dimerization of acetylene with a Nieuwland catalyst in aqueous media. *J. Phys. Org. Chem.* **2008**, *21*, 510–515. [CrossRef]

9. Liu, H.Y.; Xie, J.W.; Liu, P.; Dai, B. Effect of Cu^+/Cu^{2+} Ratio on the Catalytic Behavior of Anhydrous Nieuwland Catalyst during Dimerization of Acetylene. *Catalysts* **2016**, *6*, 120. [CrossRef]

10. Liu, J.G.; Han, M.H.; Wang, Z.W. Studies on the catalytic performance of the Nieuwland catalyst and anhydrous catalyst in the dimerization of acetylene to monovinylacetylene. *Adv. Mater. Res.* **2012**, *550–553*, 312–316. [CrossRef]

11. Liu, J.G.; Han, M.H.; Wang, Z.W. Effect of solvent on catalytic performance of anhydrous catalyst in acetylene dimerization to monovinylacetylene. *J. Energy Chem.* **2013**, *22*, 599–604. [CrossRef]

12. Liu, H.Y.; Xie, J.W.; Liu, P.; Liu, Z.Y.; Dai, B. Study on anhydrous catalyst in acetylene dimerization to monovinylacetylene. *J. Shihezi Univ.* **2015**, *33*, 622–627. [CrossRef]

13. Han, M.H.; Liu, J.G.; Zuo, Y.Z.; Wang, Z.W. Nieuwland Catalyst Preparation Method and Application. CN Patent 103285931 A, 27 February 2012.

14. Han, M.H.; Liu, J.G.; Zuo, Y.Z.; Wang, Z.W. Monovinylacetylene Product and Preparation Method of Monovinylacetylene. CN Patent 103288572 A, 11 September 2013.

15. Tao, C.Y.; Du, J.; Fan, X.; Liu, Z.H.; Xiao, C.C.; Ma, J.X.; Xue, H.T.; Shen, H.N.; Sun, D.G.; Liu, R.L. A Method of Preparing Hydrous Catalyst of Synthesizing Monovinylacetylene. CN Patent 101940953 A, 12 January 2011.

16. Tao, C.Y.; Du, J.; Fan, X.; Liu, Z.H.; Xiao, C.C.; Ma, J.X.; Xue, H.T.; Shen, H.N.; Sun, D.G.; Liu, R.L. An Improved Water System Nieuwland Catalyst and Its Application. CN Patent 101733150 A, 18 December 2009.

17. Tao, C.Y.; Du, J.; Fan, X.; Liu, Z.H.; Sun, D.G.; Shen, H.N.; Liu, R.L.; Zuo, Z.H.; Xiao, C.C.; Zhen, X.X. A Method of Preparing Catalyst of Synthesizing Monovinylacetylene and Application. CN Patent 101786022 A, 28 July 2010.

18. Liu, J.G.; Zuo, Y.Z.; Han, M.H.; Wang, Z.W.; Wang, D.Z. Stability improvement of the Nieuwland catalyst in the dimerization of acetylene to monovinylacetylene. *J. Nat. Gas Chem.* **2012**, *21*, 495–500. [CrossRef]

19. Liu, J.G.; Zuo, Y.Z.; Han, M.H.; Wang, Z.W. Improvement of anhydrous catalyst stability in acetylene dimerization by regulating acidity. *J. Chem. Technol. Biotechnol.* **2013**, *88*, 408–414. [CrossRef]

20. Liu, Z.H.; Yu, Y.L.; Tao, C.Y.; Du, J.; Liu, R.L.; Fan, X.; Peng, M.; Sun, D.G.; Zuo, Z.H.; Ma, J.X. A Method of Dimerization of Acetylene to Monovinylacetylene. CN Patent 102775266 A, 14 November 2012.

21. Han, M.H.; Liu, J.G.; Zuo, Y.Z.; Wang, Z.W. The Method and Application of Anhydrous Nieuwland Catalyst during Dimerization of Acetylene. CN Patent 103285925 A, 11 September 2013.

22. Liu, Z.H.; Yu, Y.L.; Tao, C.Y.; Fan, X. Acetylene dimerization catalyzed by LaCl$_3$-modified Nieuwland catalyst. *CIESC J.* **2014**, *65*, 1260–1266. [CrossRef]

23. Liu, Z.H.; Yu, Y.L.; Tao, C.Y.; Du, J.; Liu, R.L.; Fan, X.; Peng, M.; Sun, D.G.; Zuo, Z.H.; Ma, J.X. A Method of Synthesizing Monovinylacetylene in Aqueous Solution. CN Patent 102731241 A, 17 October 2012.

24. Du, J.; Liu, Z.H.; Tao, C.Y. A Method of Synthesizing Monovinylacetylene. CN Patent 102964198 A, 13 March 2013.

25. Zhang, Y.K.; Jia, Z.K.; Zhen, B.; Han, M.H. Investigation of the acetylene dimerization catalyzed by Nieuwland catalyst. *CIESC J.* **2016**, *67*, 294–299. [CrossRef]

26. Lu, J.L.; Xie, J.W.; Liu, H.Y.; Liu, P.; Liu, Z.Y.; Dai, B. Strontium chloride modified Nieuwland catalyst in the dimerization of acetylene to monovinylacetylene. *Asian J. Chem.* **2014**, *26*, 8211–8214. [CrossRef]

27. Lu, J.L.; Liu, H.Y.; Xie, J.W.; Liu, P.; Liu, Z.Y.; Dai, B. Study on catalytic activity of zinc(II)-copper(I) collaborative bimetallic catalysis in acetylene dimerization reaction. *J. Shihezi Univ.* **2014**, *32*, 213–217. [CrossRef]

28. Lu, J.L.; Liu, H.Y.; Xie, J.W.; Liu, P.; Dai, B.; Liu, Z.Y. Effect of polyethylene glycol/Nieuwland catalyst on acetylene dimerization reaction. *Chem. Eng.* **2015**, *43*, 60–64. [CrossRef]

 *catalysts* 

Article

Ruthenium(II)-Arene Complexes of the Water-Soluble Ligand CAP as Catalysts for Homogeneous Transfer Hydrogenations in Aqueous Phase [†]

Antonella Guerriero [1], Maurizio Peruzzini [1,2,*] and Luca Gonsalvi [1,*]

[1]   Consiglio Nazionale delle Ricerche (CNR), Istituto di Chimica dei Composti OrganoMetallici (ICCOM), Via Madonna del Piano 10, 50019 Sesto Fiorentino (Florence), Italy; antonella.guerriero@iccom.cnr.it

[2]   Consiglio Nazionale delle Ricerche (CNR), Dipartimento di Scienze Chimiche e Tecnologia dei Materiali (DSCTM), Via dei Taurini 19, 00185 Rome, Italy

*   Correspondence: maurizio.peruzzini@cnr.it (M.P.); l.gonsalvi@iccom.cnr.it (L.G.); Tel.: +39-06-49937833 (M.P.); +39-055-5225251 (L.G.)

†   Dedicated to Prof. Peter M. Maitlis on occasion of his 85th birthday.

Received: 29 December 2017; Accepted: 15 February 2018; Published: 22 February 2018

Abstract: The neutral Ru(II) complex κP-[RuCl$_2$(η^6-p-cymene)(CAP)] (**1**), and the two ionic complexes κP-[RuCl(η^6-p-cymene)(MeCN)(CAP)]PF$_6$ (**2**) and κP-[RuCl(η^6-p-cymene)(CAP)$_2$]PF$_6$ (**3**), containing the water-soluble phosphine 1,4,7-triaza-9-phosphatricyclo[5.3.2.1]tridecane (CAP), were tested as catalysts for homogeneous hydrogenation of benzylidene acetone, selectively producing the saturated ketone as product. The catalytic tests were carried out in aqueous phase under transfer hydrogenation conditions, at mild temperatures using sodium formate as hydrogen source. Complex **3**, which showed the highest stability under the reaction conditions applied, was also tested for C=N bond reduction from selected cyclic imines. Preliminary NMR studies run under pseudo-catalytic conditions starting from **3** showed the formation of κP-[RuH(η^6-p-cymene)(CAP)$_2$]PF$_6$ (**4**) as the pivotal species in catalysis.

Keywords: 1,4,7-triaza-9-phosphatricyclo[5.3.2.1]tridecane; water-soluble phosphines; ruthenium arene complexes; ketone and imine transfer hydrogenation; aqueous phase catalysis

---

1. Introduction

The importance of ruthenium compounds as homogeneous catalysts for the synthesis of fine chemicals and pharmaceutical intermediates has been widely demonstrated [1–5]. Some of them display high reactivity and selectivity in many catalytic transformations and, in particular, in the hydrogenation of different unsaturated compounds, such as α,β-unsaturated ketones and imines, using molecular H$_2$ and transfer hydrogenation (TH) protocols as reducing agents [6–9]. The employment of these last methodologies offers attractive advantages including the use of inexpensive and easy to handle hydrogen donors instead of explosive hydrogen gas, mild reaction conditions, and the possibility of using environmentally friendly solvents such as water [10–12]. The development of metal complexes containing water soluble ligands is the most common strategy to perform catalytic hydrogenations in water or under aqueous biphasic conditions and many organophosphines have been prepared for this purpose [13,14].

The water soluble phosphine 1,3,5-triaza-7-phospaadamantane (PTA, Figure 1) and its derivatives found large use as ligands for transition metal complexes active in homogeneous aqueous phase and biphasic catalysis and in the course of last decade many reports concerning Ru-PTA catalysts appeared in the literature [15–17]. For example, cyclopentadienyl (Cp) and pentamethylcyclopentadienyl (Cp*) Ru(II) complexes such as [RuCpCl(PTA)$_2$], [RuCp(MeCN)(PTA)$_2$](PF$_6$), [RuCp*Cl(PTA)$_2$], and [RuCp*(MeCN)(PTA)$_2$](PF$_6$) were proven to be active catalysts for the hydrogenation of benzylidene acetone (BZA) under H$_2$ pressure in a biphasic water/octane solvent mixture showing a high chemoselectivity to C=C double bond reduction [18,19]. Another class of catalytically active compounds is represented by [RuCl$_2$(η^6-p-cymene)(PTA)] (RAPTA-C) and [RuCl(η^6-p-cymene)(PTA)$_2$](BF$_4$), which were shown to be active catalysts for the full hydrogenation of various substituted arenes into the corresponding cyclohexanes under biphasic conditions [20]. Some Ru(II) complexes bearing 'upper rim' PTA derivatives—i.e., pending arms on the C atom adjacent to the P donor (Figure 1)—were also tested in catalytic hydrogenation of acetophenone giving good conversions even at room temperature using protocols based on the presence of both H$_2$ gas and ^{t}BuOK/^{i}PrOH [21].

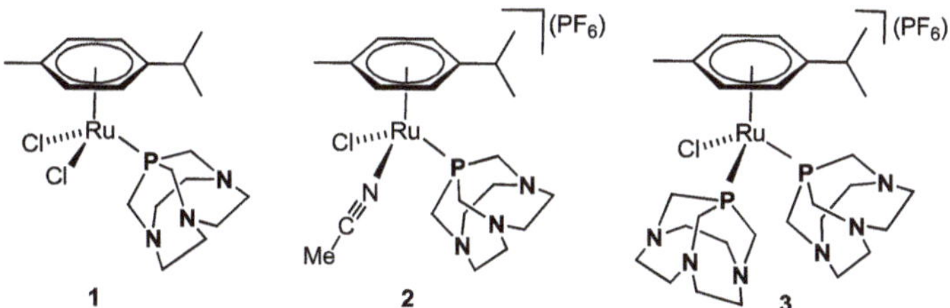

Figure 1. PTA ligand and Ru(II)-arene complexes bearing 'upper rim' PTA derivatives.

Recently, we reported on the synthesis of a new class of half-sandwich ruthenium arene complexes bearing a higher homologue of PTA, namely 1,4,7-triaza-9-phosphatricyclo[5.3.2.1]tridecane, CAP [22]. As PTA, this compound is a white, air-stable solid under standard conditions, and it is moderately water soluble with $S(H_2O)_{25\,°C} = 20$ g L^{-1}, about one order of magnitude lower than for PTA. Although structurally very similar to PTA, CAP has a higher cage flexibility due to the presence of two CH$_2$ spacers between the N atoms instead of one [23,24]. The Ru(II) complex κP-[RuCl$_2$(η^6-p-cymene)(CAP)] (**1**, Figure 2) was obtained by reacting [RuCl$_2$(η^6-p-cymene)]$_2$ with 2 equiv. of CAP in CH$_2$Cl$_2$ at room temperature. The two cationic complexes κP-[RuCl(η^6-p-cymene)(MeCN)(CAP)](PF$_6$) (**2**) and κP-[RuCl(η^6-p-cymene)(CAP)$_2$](PF$_6$) (**3**, Figure 2) were synthesized by chloride abstraction from **1** with TlPF$_6$ either in the presence of MeCN or a second equivalent of CAP, respectively. These compounds were tested in vitro for their cytotoxic activity against selected cancer cell lines with good results [22].

Figure 2. Ru(II)-arene complexes **1**–**3** bearing ligand CAP [22].

Based on the good catalytic properties shown by Ru(II) arene PTA complexes, we tested complexes **1**–**3** as catalysts for the reduction of α,β-unsaturated compounds by a mild TH protocol (HCOONa/H$_2$O/MeOH), choosing BZA as model substrate. Complex **3** was also tested under the same conditions for the C=N bond hydrogenation of selected cyclic imines to produce the corresponding amines.

Finally, in order to clarify the nature of catalytically active species involved in the hydrogenation process, some preliminary mechanistic NMR experiments under pseudo-catalytic conditions were run starting from 3 and the results are presented below.

2. Results

2.1. Catalytic Transfer Hydrogenation of BZA

As discussed above, several Ru(II) complexes including PTA and derivatives have been used as homogeneous catalysts for hydrogenation of a wide variety of unsaturated substrates [15–17]. We selected benzylidene acetone (BZA) as model substrate for α,β-unsaturated ketones hydrogenation (Scheme 1) and tested compounds **1–3** as catalysts using a TH protocol involving HCOONa/H$_2$O with a standard catalyst/substrate/formate ratio of 1:100:1000.

Scheme 1. Hydrogenation of benzylidene acetone (BZA).

Sodium formate was chosen as it is one of the mildest reducing agents with large compatibility with various functional groups, also having good water solubility. To facilitate the formation of a homogeneous liquid phase and to ensure complete solubility of all reagents, the tests were carried out in a water/methanol 1:1 mixture. The results are summarized in Table 1.

Table 1. Catalytic transfer hydrogenation of benzylidene acetone (BZA) with complexes **1–3** [a].

Entry	Catalyst	T (°C)	% Conv.	Time (h)	Yield A (%) [b]	Yield B (%) [b]	Yield C (%) [b]
1	1	60	60.0	24	45.3	13.0	1.7
2 [c]	1	60	18.7	24	14.2	4.4	0.1
3	1	80	99.4	4	82.0	5.2	12.2
4 [c]	1	80	76.3	4	61.7	12.8	1.8
5	2	80	82.6	4	68.9	11.5	2.2
6 [c]	2	80	51.9	4	40.4	10.8	0.7
7	3	80	71.8	24	52.3	15.8	3.7
			96.0	48	68.7	13.5	13.8
8 [c]	3	80	66.5	24	46.6	17.8	2.1
			92.5	48	73.5	13.0	6.0
9	RAPTA-C	80	99.5	24	93.2	3.2	3.1
10 [c]	RAPTA-C	80	30.0	24	23.8	6.0	0.2

[a] General conditions: catalyst, 9.8×10^{-3} mmol; BZA, 0.98 mmol; HCOONa, 9.8 mmol; MeOH:H$_2$O (1:/1), 6 mL. [b] GC values based on pure samples, except B (see Experimental Section): A = 4-phenyl-2-butanone; B = 4-phenyl-3-buten-2-ol; C = 4-phenyl-2-butanol. [c] Hg(0) added (one drop).

At first, the catalytic run was carried out at 60 °C in the presence of complex **1** (entry 1) showing a conversion of 60% after 24 h and a ca. 3.5:1 ratio between the saturated ketone (A) and the unsaturated alcohol (B). When the temperature was raised to 80 °C (entry 3), the conversion reached 99.4% after 4 h only. A high conversion of 82.6% was also observed with **2** after 4 h reaction under the same conditions (entry 5), also in this case with high selectivity (83.4%) towards the saturated ketone. However, darkening of the reaction mixture and the appearance of a black precipitate during tests run either at 60 °C or 80 °C with both catalysts **1** and **2** suggested that these complexes were at least partially decomposing under these conditions. This was confirmed by repeating the runs using the standard Hg(0) poisoning test which resulted in the reduction of the catalytic activity. This was particularly evident for complex **2**, with an important decrease of conversion after 4 h at 80 °C (51.9%, entry 6, vs. 82.6%, entry 5).

For comparison, we tested complex RAPTA-C, i.e., the PTA analogue of **1**, under the same reaction conditions as in Table 1, entry 3. Also in this case a substantial drop of conversion was observed in the presence of Hg (30%, entry 10, vs. 99.5% without Hg, entry 9), indicating that a significant degree of decomposition leading to undefined catalytically active phosphine-free species was probably occurring also in the case of RAPTA-C.

Complex **3** was then tested and showed to be more stable than **1** and **2** under the catalytic conditions applied as no black precipitate was observed during the catalytic runs even after 48 h at 80 °C. The catalytic activity was only slightly reduced in the presence of Hg(0), giving 92.5% BZA conversion after 48 h at 80 °C (entry 8) compared to the 96.0% achieved in the test run without addition of mercury (entry 7). The higher stability was accompanied by a lower reaction rate to have almost complete conversion that was reached only after 48 h at 80 °C. On the other hand, the choice of this higher temperature resulted to be essential to achieve good activity with **3**, as only 7.7% BZA conversion was obtained after 24 h in tests run at 60 °C.

As with **1** and **2**, complex **3** also proved to be a rather chemoselective catalyst for C=C bond hydrogenation, converting BZA mainly to the saturated ketone 4-phenyl-butan-2-one (A) with a selectivity of 71.5% at 92.5% conversion after 48 h.

As an example of cyclic α,β-unsaturated ketone, 2-cyclohexen-1-one was used as substrate. In the case of complex **1**, after 4 h at 80 °C using the standard conditions for TH described above, the almost complete conversion was observed. However, mainly cyclohexanol (62.9%) was obtained, demonstrating poor chemoselectivity to either C=C or C=O bond reduction. Moreover, a non-negligible amount of cyclohexanone dimethyl ketal (ca. 36.6%) was identified in solution by GC-MS at the end of the run, suggesting that C=O bond protection was competing with hydrogenation under these conditions. When the tests were performed with catalyst **3**, low yields (ca. 1%) of hydrogenation products were observed after 24 h at 80 °C. The formation of the corresponding ketal was again observed (ca. 27%).

2.2. Transfer Hydrogenation of Cyclic Imines

Hydrogenation of imines and, in particular, cyclic imines is a worthy synthetic tool to obtain amines, that in turn are synthetic intermediates of great importance especially in pharmaceutical industry [25]. Among Ru catalysts described in the literature for this class of homogeneously catalyzed reaction, outstanding performances were obtained for example with Noyori's complex [RuCl(η^6-*p*-cymene)(TsDPEN)] (TsDPEN=*N*-(*p*-toluene-sulfonyl)-1,2-diphenylethylenediamine), using a formic acid-triethylamine azeotropic mixture as reducing agent [26,27]. Examples of water soluble ruthenium–arene catalysts for TH of imines in aqueous phase with HCOONa are also known [28,29]. For cyclic imines hydrogenation, we also described an efficient system based on a heterogenized Ir(I) complex bearing PTA supported into an ion-exchange resin, using water and H_2 pressure [30].

On the basis of the higher stability demonstrated by complex **3** compared to **1** and **2** under the catalytic conditions previously discussed, this compound was selected as a catalyst for tests in TH of the cyclic imines shown in Table 2.

Table 2. Catalytic transfer hydrogenation of cyclic imines with **3** [a].

Entry	T (°C)	Solvent	Substrate	Product [b]	Yield [c]	Time (h)
1	80	H_2O	3,4-dihydroisoquinoline	1,2,3,4-tetrahydroisoquinoline	39.9	24
2	60	MeOH/H_2O 1:1			10.9	24
3	80	MeOH/H_2O 1:1			74.1 92.5	24 48

Table 2. *Cont.*

Entry	T (°C)	Solvent	Substrate	Product [b]	Yield [c]	Time (h)
4	80	MeOH/H$_2$O 1:1	2-methylquinoxaline	2-methyl-1,2,3,4-tetrahydroquinoxaline	0.4	24
5	80	MeOH/H$_2$O 1:1	5-methylquinoxaline	5-methyl-1,2,3,4-tetrahydroquinoxaline	0.0	48
6	80	MeOH/H$_2$O 1:1	quinoline	1,2,3,4-tetrahydroquinoline	1.5 / 3.0	24 / 48

[a] General conditions: catalyst, 1.0×10^{-2} mmol; substrate, 1.0 mmol; HCOONa, 10.0 mmol; solvent, 6 mL. [b] Products confirmed by GC and GC-MS analyses. [c] GC values based on pure samples where available.

The catalytic tests were at first performed with 3,4-dihydroisoquinoline to obtain selectively 1,2,3,4-tetrahydroisoquinoline, a moiety often present in pharmaceutical drugs [31]. As for TH of BZA, a catalyst/substrate/HCOONa ratios of 1:100:1000 were used. The first experiment was carried out in neat water, but as in the case of BZA a MeOH/H$_2$O (1:1) mixture was required to guarantee the formation of a homogeneous phase. In fact, a conversion of only ca. 40% was obtained after 24 h at 80 °C in neat water (Table 2, entry 1). On the other hand, when the reaction was carried out in MeOH/H$_2$O (1:1) mixture, the conversion to 1,2,3,4-tetrahydroisoquinoline reached 74.1% after 24 h and 92.5% after 48 h at 80 °C (Table 2, entry 3). As already noted for BZA, a temperature of 80 °C was required to achieve high conversions, as demonstrated by the test performed at 60 °C (Table 2, entry 2) that gave only ca. 11% conversion after 24 h.

The optimized test conditions were used with **3** and three cyclic imines, namely 2-methylquinoxaline, 5-methylquinoxaline, and quinoline (Table 2, entries 4–6). Disappointingly, these tests showed only minor or no substrate conversion even after 48 h at 80 °C.

As an example of acyclic imine reduction, *N*-(1-phenylethylidene)aniline was tested with **3** under the conditions described above (see Table 2 caption), but in this case extensive substrate decomposition was observed, in agreement with previous studies with Ir-PTA catalysts [30].

2.3. NMR Mechanistic Studies under Pseudo-Catalytic Conditions

Preliminary mechanistic studies were carried out by NMR spectroscopy with complex **3** under pseudo-catalytic conditions. Firstly, 20 equiv. of HCOONa were added to compound **3** dissolved in MeOH and the resulting orange solution was immediately analyzed by ^{31}P{^{1}H} NMR. The spectrum showed a singlet at 52.56 ppm, corresponding to unreacted **3**, and an accompanying septet at -143.76 ppm due to the PF$_6$ anion. No changes were observed after heating the solution to 60 °C for 2 h; thus, an additional 30 equiv. of HCOONa were added to the reaction mixture and the NMR tube was left standing at 60 °C for 24 h. After this time, the ^{31}P{^{1}H} NMR spectrum showed complete conversion of **3** to a new species (**4**) characterized by a singlet at 74.34 ppm and a septet at -143.79 ppm (PF$_6$). The ^{1}H NMR spectrum gave a triplet in the negative region at -11.43 ppm ($^{2}J_{HP} = 35.3$ Hz), as expected for the formation of a Ru-H bond. In another test, **3** was reacted directly with 50 equiv. of HCOONa in MeOH/H$_2$O (1:1) at 80 °C. After 3 h, in addition to the singlet at 55.77 ppm and the septet at -144.45 ppm due to **3** (slight differences in the chemical shift values derive from the use of MeOH/H$_2$O mixture instead of pure MeOH), the singlet due to **4** was detected in the ^{31}P{^{1}H} NMR spectrum at 78.26 ppm, in 4:1 ratio with **3**. Complete conversion of **3** to **4** was reached after leaving the tube standing at 80 °C for 17 h. Upon solvent removal in vacuo, and dissolving the obtained solid in CD_2Cl_2, the $^{31}P\{^1H\}$ NMR spectrum of **4** showed a singlet at 67.66 ppm and a septet at −149.50 ($^1J_{PF}$ = 711 Hz) ppm due to PF_6. The corresponding ^1H NMR spectrum showed in the negative region a triplet at −12.23 ppm ($^2J_{HP}$ = 35.5 Hz) which became a singlet in the corresponding P-decoupled ^1H $\{^{31}P\}$ NMR spectrum (Figure 3).

Based on this evidence and by comparison with data reported for [RuCp(H)(PTA)$_2$], whose hydride signal was identified by ^1H NMR as a triplet at −14.36 ppm ($^2J_{HP}$ = 36.6 Hz) in CD_2Cl_2 [32], we attribute the new pattern to the formation of the cationic monohydrido complex κP-[RuH(η^6-p-cymene)(CAP)$_2$](PF$_6$) (**4**, Scheme 2). A confirmation of this attribution came from the independent synthesis of **4** (see Experimental Section).

A final NMR scale experiment was carried out with **3** and HCOONa (1:50 ratio) under the same conditions of solvents (MeOH/H$_2$O 1:1) and temperature (80 °C) described above, adding 3,4-dihydroisoquinoline (5 equiv. to **3**) to the mixture before heating. The pattern due to **4** was observed initially after 4 h, to become the only P-containing species after 24 h. The conversion of the imine to the corresponding amine 1,2,3,4-tetrahydroisoquinoline was confirmed by GC-MS analysis of the crude mixture after 24 h.

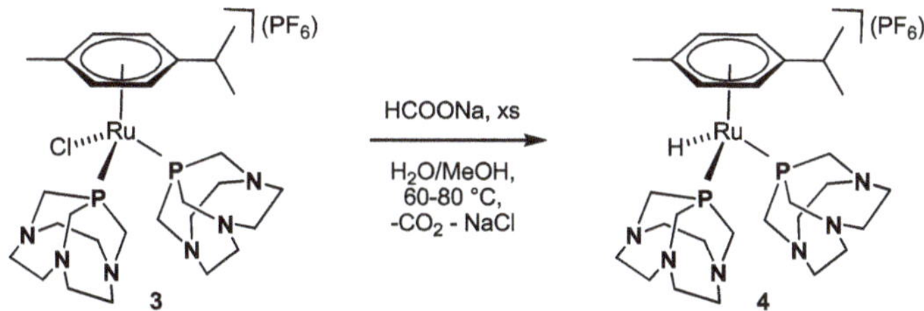

Scheme 2. Conversion of **3** to **4** by reaction with HCOONa in H$_2$O/MeOH.

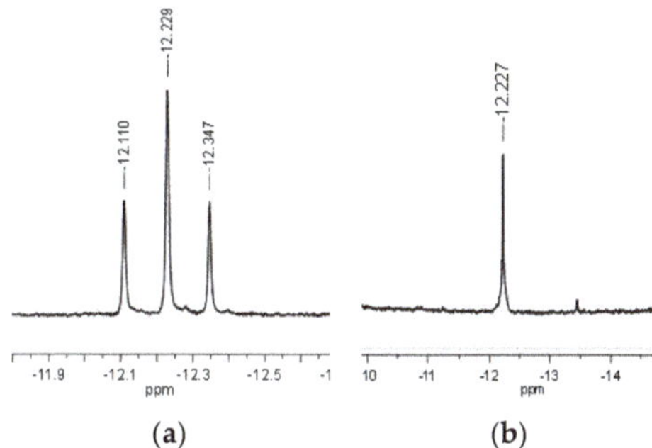

Figure 3. ^1H (**a**) and ^1H $\{^{31}P\}$ NMR (**b**) spectra (negative region only, CD$_2$Cl$_2$) showing the change from triplet to singlet for the Ru-H signal in **4**.

3. Experimental Section

3.1. Materials and Methods

All manipulations were carried out under a purified N$_2$ atmosphere using standard Schlenk techniques unless otherwise noted. Deuterated solvents and other reagents were bought from commercial suppliers and used without further purification. Doubly distilled water was used. All solvents were distilled, dried, and degassed prior to use. Compounds κP-[RuCl$_2$(η^6-p-cymene)(CAP)] (**1**), κP-[RuCl(η^6-p-cymene)(MeCN)(CAP)](PF$_6$) (**2**), κP-[RuCl(η^6-p-cymene)(CAP)$_2$](PF$_6$) (**3**) [22], RAPTA-C [33] and [RuCl$_2$(η^6-p-cymene)]$_2$ [34] were prepared as described in the literature. ^1H and

$^{31}P\{^1H\}$ NMR spectra were recorded on a Bruker DRX300 spectrometer (operating at 300.13 and 121.50 MHz, respectively). The ^{31}P spectra were normally run with proton decoupling and are reported in ppm relative to an external H_3PO_4 standard, with downfield positive shifts. For catalytic tests, all substrates were bought from Aldrich and used without further purification, with the exception of benzylidene acetone which was recrystallized from hot toluene. 3,4-dihydroisoquinoline and BZA together with their reduction products—namely 1,2,3,4-tetrahydroisoquinoline (95%, Sigma-Aldrich S.r.l., Milan, Italy), 4-phenyl-2-butanone (98%, Sigma-Aldrich S.r.l., Milan, Italy), and 4-phenyl-2-butanol (97%, Sigma-Aldrich S.r.l., Milan, Italy,)—were used as standards for GC analyses; on the contrary, compound 4-phenyl-3-buten-2-ol has been identified by GC-MS: m/z (%) 148 [M]$^+$ (50), 129 (100) [35]. All GC analyses were performed on a Shimadzu GC 2010 Plus (Shimadzu _Italia S.r.l., Milan, Italy) gas chromatograph (carrier gas: He; injection mode: split at 250 °C,) equipped with flame ionization detector and a Supelco (part of Sigma-Aldrich Inc., St. Louis, MO, USA) SPB^{TM}-1 capillary column (30 m, 0.25 mm ID, 0.25 μm film thickness). In the case of hydrogenation of BZA, the GC method started from a column temperature of 110 °C (hold time: 12 min) to increase 12 °C/min up to 240 °C (hold time: 5 min); The initial pressure was 111.8 kPa and the split ratio 80.0; the linear velocity was set at 30.0 cm/s. All products were identified at different retention times (rt) as indicated here in detail: 4-phenyl-2-butanone, rt = 8.38 min; 4-phenyl-2-butanol, rt = 9.21 min; 4-phenyl-3-buten-2-ol, rt = 12.26 min; BZA, rt = 13.54 min. In the case of 3,4-dihydroisoquinoline, the GC method started from a column temperature of 115 °C (hold time: 12 min) to increase 5 °C/min up to 250 °C (hold time: 5 min); the initial pressure was 120.8 kPa and the split ratio 80.0; the linear velocity was set at 32.0 cm/sec. Compound 3,4-dihydroisoquinoline and 1,2,3,4-tetrahydroisoquinoline showed retention times of 6.72 min and 7.52 min, respectively.

GC-MS analyses were performed on a Shimadzu GCMS-QP2010S apparatus (Shimadzu _Italia S.r.l., Milan, Italy) equipped with a flame ionization detector and a Supelco (part of Sigma-Aldrich Inc., St. Louis, MO, USA) SPB^{TM}-1 fused silica capillary column (30 m, 0.25 mm ID, 0.25 μm film thickness), mass analyzer metal quadrupole mass filter with pre-rod, ionisation mode EI, split 250 °C, ion source temperature 200 °C, and interface temperature 280 °C. The specific analysis parameters where chosen to match those used for the corresponding GC method.

3.2. Transfer Hydrogenation Tests

All reactions were carried out in a Schlenk tube under an inert atmosphere of nitrogen using degassed solvents. In a typical experiment, HCOONa was dissolved in 2.0 mL of water and added by syringe to the solution of the catalyst in MeOH/H_2O (1:1, 2.0 mL). The temperature was set and when reached, the solution of the substrate in MeOH (2.0 mL) was added to the reaction mixture. Both during the sampling and at the end of the catalytic runs, an aliquot of the reaction mixture (0.1 mL) was taken by syringe, diluted with methanol (0.4 mL), and analyzed by GC. The products of the tests were confirmed by GC-MS analyses. Each catalytic test was repeated at least twice to check for reproducibility.

3.3. NMR Scale Experiments

In a first experiment, a Schlenk tube was charged under an inert atmosphere of nitrogen with complex **3** (4.0 mg, 0.005 mmol) and HCOONa (17.0 mg, 0.25 mmol), which were dissolved in 1.0 mL of degassed solvent (MeOH or MeOH/H_2O 1:1 mixture). Then, 0.7 mL of the resulting clear solution was trasferred by syringe into a NMR tube containing a C_6D_6 capillary for deuterium lock. In a second experiment, the same procedure was repeated, also adding solid 3,4-dihydroisoquinoline (3.3 mg, 0.025 mmol) to the solution before heating. In both cases, the solutions were first analyzed at room temperature, then the NMR tubes were placed in a oil bath set at the desired temperature, taking ^1H and $^{31}P\{^1H\}$ NMR spectra at room temperature and different intervals of time to monitor the course of the reactions.

3.4. Synthesis of κP-[RuH(η^6-p-cymene)(CAP)$_2$](PF$_6$) **4**

A Schlenk tube was charged under an inert atmosphere of nitrogen with complex **3** (30 mg, 0.037 mmol) and HCOONa (125.3 mg, 1.84 mmol), adding 8 mL of degassed MeOH/H$_2$O (1:1 mixture). The resulting orange mixture was left stirring at 80 °C for 28 h. After this time, the solvent was removed in vacuo and the solid residue dissolved in CH$_2$Cl$_2$. After filtration, to the resulting solution, diethylether was added to precipitate the product as a light orange solid, that was finally dried under vacuum (isolated yield ca. 50%).

Analysis: ^{1}H NMR, negative region: δ(ppm, CD$_2$Cl$_2$) −12.23 ppm ($^2J_{HP}$ = 35.5 Hz); ^{31}P{^{1}H} NMR: δ(ppm, CD$_2$Cl$_2$) 67.51 (s); −149.53 (sept, $^1J_{PF}$ = 711.1 Hz, PF$_6$).

4. Conclusions

The half-sandwich ruthenium(II) arene complexes **1–3** bearing the water soluble ligand CAP were tested in homogeneous catalytic transfer hydrogenations of some selected unsaturated substrates. Whereas the monophosphine complexes **1** and **2** showed poor overall stability under the reaction conditions applied for the tests, the bisphosphine complex **3** proved to be more stable and active in the hydrogenation of benzylidene acetone and 3,4-dihydroisoquinoline using a very mild reduction protocol such as transfer hydrogenation with sodium formate. NMR studies in solution performed with **3** under pseudo-catalytic conditions clearly showed the formation of the monohydride derivative **4**, which is likely the active form of the catalyst during the reaction.

Acknowledgments: The Italian Ministry for Education and Research (MIUR) is kindly acknowledged for financial support through Project PRIN 2015 (grant number 20154X9ATP).

Author Contributions: A.G. and L.G. conceived and designed the experiments; A.G. performed the experiments and analyzed the data; M.P. contributed with all the authors to write and approve the final version of the paper.

Conflicts of Interest: The authors declare no conflict of interest.

References

1. Morris, R.H. Ruthenium and Osmium. In *The Handbook of Homogeneous Hydrogenation*; de Vries, J.G., Elsevier, C.J., Eds.; WILEY-VCH Verlag GmbH & Co. KGaA: Weinheim, Germany, 2007; Chapter 3, pp. 45–70. ISBN 978-3-527-31161-3.
2. Clapham, S.E.; Hadzovic, A.; Morris, R.H. Mechanisms of the H$_2$-hydrogenation and transfer hydrogenation of polar bonds catalyzed by ruthenium hydride complexes. *Coord. Chem. Rev.* **2004**, *248*, 2201–2237. [CrossRef]
3. Noyori, R.; Hashiguchi, S. Asymmetric transfer hydrogenation catalyzed by chiral ruthenium complexes. *Acc. Chem. Res.* **1997**, *30*, 97–102. [CrossRef]
4. Evans, D.; Osborn, J.A.; Jardine, F.H.; Wilkinson, G. Homogeneous Hydrogenation and Hydroformylation using Ruthenium Complexes. *Nature* **1965**, *208*, 1203–1204. [CrossRef]
5. Blaser, H.U.; Malan, C.; Pugin, B.; Spindler, F.; Steiner, H.; Studer, M. Selective Hydrogenation for Fine Chemicals: Recent Trends and New Developments. *Adv. Synth. Catal.* **2003**, *345*, 103–151. [CrossRef]
6. Zanotti-Gerosa, A.; Hems, W.; Groarke, M.; Hancock, F. Ruthenium-Catalysed Asymmetric Reduction of Ketones. *Platin. Met. Rev.* **2005**, *49*, 158–165. [CrossRef]
7. Clarke, M.L.; Dìaz-Valenzuela, M.B.; Slawin, A.M.Z. Hydrogenation of Aldehydes, Esters, Imines, and Ketones Catalyzed by a Ruthenium Complex of a Chiral Tridentate Ligand. *Organometallics* **2007**, *26*, 16–19. [CrossRef]
8. James, B.R. Synthesis of chiral amines catalyzed homogeneously by metal complexes. *Catal. Today* **1997**, *37*, 209–221. [CrossRef]
9. Samec, J.S.M.; Bäckvall, J.E. Ruthenium-Catalyzed Transfer Hydrogenation of Imines by Propan-2-ol in Benzene. *Chem. Eur. J.* **2002**, *8*, 2955–2961. [CrossRef]
10. Foubelo, F.; Nájera, C.; Yus, M. Catalytic asymmetric transfer hydrogenation of ketones: Recent advances. *Tetrahedron Asymmetry* **2015**, *26*, 769–790. [CrossRef]
11. Ito, J.-I.; Nishiyama, H. Recent topics of transfer hydrogenation. *Tetrahedron Lett.* **2014**, *55*, 3133–3146. [CrossRef]

12. Wang, D.; Astruc, D. The Golden Age of Transfer Hydrogenation. *Chem. Rev.* **2015**, *115*, 6621–6686. [CrossRef] [PubMed]

13. Shaughnessy, K.H. Hydrophilic ligands and their application in aqueous-phase metal-catalyzed reactions. *Chem. Rev.* **2009**, *109*, 643–710. [CrossRef] [PubMed]

14. James, B.R.; Lorenzini, F. Developments in the chemistry of tris(hydroxymethyl)phosphine. *Coord. Chem. Rev.* **2010**, *254*, 420–430. [CrossRef]

15. Phillips, A.D.; Gonsalvi, L.; Romerosa, A.; Vizza, F.; Peruzzini, M. Coordination chemistry of 1,3,5-triaza-7-phospaadamantane (PTA). Transition metal complexes and related catalytic, medicinal and photoluminescent applications. *Coord. Chem. Rev.* **2004**, *248*, 955–993. [CrossRef]

16. Bravo, J.; Bolaño, S.; Gonsalvi, L.; Peruzzini, M. Coordination chemistry of 1,3,5-triaza-7-phospaadamantane (PTA) and derivatives. Part II. The quest for tailored ligands, complexes and related applications. *Coord. Chem. Rev.* **2010**, *254*, 555–607. [CrossRef]

17. Guerriero, A.; Peruzzini, M.; Gonsalvi, L. Coordination chemistry of 1,3,5-triaza-7-phospaadamantane (PTA) and derivatives. Part III. Variations on a theme: Novel architectures, materials and applications. *Coord. Chem. Rev.* **2018**, *355*, 328–361. [CrossRef]

18. Akbayeva, D.N.; Gonsalvi, L.; Oberhauser, W.; Peruzzini, M.; Vizza, F.; Brüggeller, P.; Romerosa, A.; Sava, G.; Bergamo, A. Synthesis, catalytic properties and biological activity of new water soluble ruthenium cyclopentadienyl PTA complexes [(C$_5$R$_5$)RuCl(PTA)$_2$] (R= H, Me; PTA= 1,3,5-triaza-7-phosphaadamantane). *Chem. Commun.* **2003**, 264–265. [CrossRef]

19. Bolaño, S.; Gonsalvi, L.; Zanobini, F.; Vizza, F.; Bertolasi, V.; Romerosa, A.; Peruzzini, M. Water soluble ruthenium cyclopentadienyl and aminocyclopentadienyl PTA complexes as catalysts for selective hydrogenation of α,β-unsaturated substrates (PTA = 1,3,5-triaza-7-phosphaadamantane). *J. Mol. Catal. A Chem.* **2004**, *224*, 61–70. [CrossRef]

20. Dyson, P.J.; Ellis, D.J.; Laurenczy, G. Minor Modifications to the Ligands Surrounding a Ruthenium Complex Lead to Major Differences in the Way in which they Catalyse the Hydrogenation of Arenes. *Adv. Synth. Catal.* **2003**, *345*, 211–215. [CrossRef]

21. Krogstad, D.A.; Guerriero, A.; Ienco, A.; Manca, G.; Peruzzini, M.; Reginato, G.; Gonsalvi, L. Imidazolyl-PTA Derivatives as Water-Soluble (P,N) Ligands for Ruthenium-Catalyzed Hydrogenations. *Organometallics* **2011**, *30*, 6292–6302. [CrossRef]

22. Guerriero, A.; Oberhauser, W.; Riedel, T.; Peruzzini, M.; Dyson, P.J.; Gonsalvi, L. New Class of Half-Sandwich Ruthenium(II) Arene Complexes Bearing the Water-Soluble CAP Ligand as an in Vitro Anticancer Agent. *Inorg. Chem.* **2017**, *56*, 5514–5518. [CrossRef] [PubMed]

23. Britvin, S.N.; Lotnyk, A. Water-Soluble Phosphine Capable of Dissolving Elemental Gold: The Missing Link between 1,3,5-triaza-7-phosphaadamantane (PTA) and Verkade's Ephemeral Ligand. *J. Am. Chem. Soc.* **2015**, *137*, 5526–5535. [CrossRef] [PubMed]

24. Britvin, S.N.; Rumyantsev, A.M.; Zobnina, A.E.; Padkina, M.V. Between Adamantane and Atrane: Intrabridgehead Interactions in the Cage-Like Phosphane Related to a Novel Tris(homoadamantane) Ring System. *Chem. Eur. J.* **2016**, *22*, 14227–14235. [CrossRef] [PubMed]

25. Fleury-Brégeot, N.; de la Fuente, V.; Castillón, S.; Claver, C. Highlights of Transition Metal-Catalyzed Asymmetric Hydrogenation of Imines. *ChemCatChem* **2010**, *2*, 1346–1371. [CrossRef]

26. Hashiguchi, S.; Fujii, A.; Takehara, J.; Ikariya, T.; Noyori, R. Asymmetric Transfer Hydrogenation of Aromatic Ketones Catalyzed by Chiral Ruthenium(II) Complexes. *J. Am. Chem. Soc.* **1995**, *117*, 7562–7563. [CrossRef]

27. Uematsu, N.; Fujii, A.; Hashiguchi, S.; Ikariya, T.; Noyori, R. Asymmetric Transfer Hydrogenation of Imines. *J. Am. Chem. Soc.* **1996**, *118*, 4916–4917. [CrossRef]

28. Wu, J.; Wang, F.; Ma, Y.; Cui, X.; Cun, L.; Zhu, J.; Deng, J.; Yu, B. Asymmetric transfer hydrogenation of imines and iminiums catalyzed by a water-soluble catalyst in water. *Chem. Commun.* **2006**, 1766–1768. [CrossRef] [PubMed]

29. Canivet, J.; Süss-Fink, G. Water-soluble arene ruthenium catalysts containing sulfonated diamine ligands for asymmetric transfer hydrogenation of α-aryl ketones and imines in aqueous solution. *Green Chem.* **2007**, *9*, 391–397. [CrossRef]

30. Barbaro, P.; Gonsalvi, L.; Guerriero, A.; Liguori, F. Facile heterogeneous catalytic hydrogenation of C=N and C=O bonds in neat water: Anchoring of water-soluble metal complexes onto ion-exchange resins. *Green Chem.* **2012**, *14*, 3211–3219. [CrossRef]

31. Antkiewicz-Michaluk, L.; Wąsik, A.; Michaluk, J. 1-Methyl-1,2,3,4-Tetrahydroisoquinoline, an Endogenous Amine with Unexpected Mechanism of Action: New Vistas of Therapeutic Application. *Neurotox. Res.* **2014**, *25*, 1–12. [CrossRef] [PubMed]
32. Frost, B.J.; Mebi, C.A. Aqueous organometallic chemistry: Synthesis, structure and reactivity of the water-soluble metal hydride CpRu(PTA)$_2$H. *Organometallics* **2004**, *23*, 5317–5323. [CrossRef]
33. Allardyce, C.S.; Dyson, P.J.; Ellis, D.J.; Heath, S.L. [Ru(η^6-*p*-cymene)Cl$_2$(pta)] (pta = 1,3,5-triaza-7-phosphatricyclo[3.3.1.1]-decane): A Water Soluble Compound That Exhibits pH Dependent DNA Binding Providing Selectivity for Diseased Cells. *Chem. Commun.* **2001**, *2*, 1396–1397. [CrossRef]
34. Bennett, M.A.; Huang, T.N.; Matheson, T.W.; Smith, A.K. (η^6-Hexamethylbenzene)Ruthenium Complexes. *Inorg. Synth.* **1982**, *21*, 74–78. [CrossRef]
35. Xu, W.; Zhou, Y.; Wang, R.; Wu, G.; Chen, P. Lithium amidoborane, a highly chemoselective reagent for the reduction of α,β-unsaturated ketones to allylic alcohols. *Org. Biomol. Chem.* **2012**, *10*, 367–371. [CrossRef] [PubMed]

MDPI

St. Alban-Anlage 66

4052 Basel

Switzerland

Tel. +41 61 683 77 34

Fax +41 61 302 89 18

www.mdpi.com

Catalysts Editorial Office

E-mail: catalysts@mdpi.com

www.mdpi.com/journal/catalysts